ingenious

ingenious

A TRUE STORY OF INVENTION, THE X PRIZE,
AND THE RACE TO REVIVE AMERICA

JASON FAGONE
FOREWORD BY TOM VOELK

Skyhorse Publishing

skyhorsepublishing.com.

ging-in-Publication Data is available on file.

as

nati Motor Works

1839-5

tes of America

for **Dana** and **Mia**

contents

part three: the ragged edge

foreword

Look around. No, really. Stop reading for five seconds and consider everything that's physical in your world. Furniture. Shoes. Appliances. Smartphones. Computers. Houses. Cars.

Every one of these items is a dream made solid by a human being.

How many of us stop to consider the enormous depth of our surroundings? This world is the culmination of imagination made real. Everything we touch was preceded by an inferior version, judged by a dissatisfied individual that believed "I can make this better."

It might be as simple as a shoelace that holds its bunny ears better. Maybe it's a communications satellite that allows us to share news with loved ones from anywhere in the world.

We are all the richer for the smallest of these efforts.

Certainly, we've all had daydreams for making it rich. Few take action because, damn it, it's a hard slog to get your hands dirty and scary to endure rejection. An individual bringing a product to market faces brutal indifference, not to mention a blizzard of standards, rules, and regulations.

So most of us shrug and leave that flash of an idea to slumber in our imagination, never to help or delight anyone else. But, a determined few risk their reputations, careers, marriages, savings, and even homes to make that spark a real thing, to create a tangible object for others to see, touch, and use.

What lies in the pages before you is not simply how to make a super fuel-efficient vehicle. This book chronicles the struggle of hundreds of men and women who believe in their heart of hearts that the automobile can be significantly improved.

Some are motivated by fortune, in this case the ten million dollars in prize money offered up by the Progressive Insurance Automotive X Prize competition. Others are driven by the confidence that they can transcend the maze of bureaucracy found in the established automotive industry and hammer out a vehicle that meets the lofty demands of the X Prize committee—a safe affordable production vehicle that exceeds 100 MPGe.

The X Prize competition began in 2007 with eighty cars and sixty-seven teams throwing their hats into the ring. In the end, three of them prevail. Author Jason Fagone focuses on four diverse efforts. There's a front-running group from California with a well-known and well-funded project. A bunch of Philadelphia teenagers led by their teachers to build two cars in an after-school club. An unlikely effort is formed by a Lynchburg, Virginia real estate developer, assembling a group that constructs four, yes, four vehicles. And in an Illinois cornfield, a dreamer, his wife, and his friends struggle to produce an electric car with a motor designed in Switzerland, manufactured in Denmark, and purchased from a Russian living in Oregon. Oh, and the batteries are from China.

It's a reminder that none of us dream alone.

This three-year odyssey is no dry tale. It's a saga of misguided beginnings, maxed-out credit cards, multifaceted engineering, midnight Hail Marys and, best of all, underdogs. It's less about the automobile and more about the human condition and the innovation that rises from it. In total, tens of millions of dollars were shelled out chasing the X Prize. More valuable is the physical and emotional struggle that few of us would ever expend. These passionate pioneers, some sitting atop homemade battery packs with more energy per square inch than dynamite, put their very lives at risk.

For those who take our world for granted, know that there's a lot to learn in these pages. The automobile is by far the most complex consumer product on the planet, fusing ancient metal foundry with bleeding edge software development. Today's Toyota Prius covers an amazing fifty-one miles on a single gallon of gasoline. X Prize participants were tasked with doubling that.

To eek out a single additional mile-per-gallon of efficiency, huge wind tunnels costing hundreds of millions of dollars help to sculpt the silhouette of modern vehicles. It might save just a fraction of an ounce of fuel on each trip taken, but consider the millions of daily excursions. Tiny adjustments become financial and environmental wins. In the case of one X Prize team, air becomes a whisper's difference between winning and losing.

Let's not forget that Jason made something admirable as well: the book you're reading. Consider the time, travel, skill, passion, and finances he poured into making the X Prize competition something we can all easily understand.

Maybe, just maybe, his tale will nudge that vague idea nestled comfortably in your brain to life. Don't be afraid to wake it up. Don't be overwhelmed by the seemingly herculean effort before you. Don't be intimidated by the prospect of failure. The brutal truth in life is that not everyone wins. Riches are not always financial. Most of the X Prize competitors find out the reward is the journey itself.

That said, go out and make the world a better place.

Tom Voelk
Seattle, 2017

introduction

This is a book about a handful of Americans who entered an unusual sort of contest in the late 2000s, during the Great Recession. The participants weren't wealthy, far from it in most cases. In the beginning they were drawn to the contest by its garish prize pot: $10 million, enough to change their lives forever. But then something else started to propel the contestants, something beyond greed. It was the crazy intangible force that makes people mess up their marriages and blow up their credit and risk everything they have because they think they've found their one chance, maybe their only chance, to show their creative and inventive powers. To prove their worth to all who ever doubted them, and to themselves.

The contest was run by an organization in Southern California, the X Prize Foundation, that promotes "the power of competition" to generate "radical breakthroughs for the benefit of humanity." Since 1996, the foundation has raised money from rich individuals and corporations to offer huge cash prizes for seemingly impossible inventions. There have been X Prizes for sending rocket planes into space, for launching robots to the Moon, for designing a Star-Trek-like "Tricorder" to diagnose disease, for cleaning up oil spills in the ocean, and for transforming harmful carbon dioxide emissions from power plants into useful products.

In 2006, the foundation announced the Automotive X Prize, a $10 million contest to create the super-efficient car of the future. I saw an article about the prize and went looking for more information.

As far as I could tell, the Automotive X Prize was like a combination of the Olympics, NASCAR, *Junkyard Wars*, and the Apollo Program. The foundation had published a rough blueprint of a car that didn't exist, a car it wanted to see in the world. Now it was daring people to build this car. The foundation didn't say *how*. That was up to the inventors. What mattered were the performance targets. To win the prize, the $10 million, a car had to travel at least one hundred miles on a single gallon of gas, or the equivalent amount of energy from an alternative source: electric batteries, hydrogen, solar, whatever. The car also had to be "production-capable" and "designed to reach the market"—no concept cars, no impractical science projects. Typical American drivers had to be able to imagine these cars sitting in their driveways. And crucially, anyone could try to win the prize, any person or group, from a company the size of Ford or BMW down to a lone man or woman in a garage. There was an appealing democracy at the root of it. You didn't need to be rich or work for an auto firm or have an engineering degree. You just needed a good idea and the will to build it.

I'm writing this introduction in 2017, ten years after the contest opened, and obviously a lot has changed since then. The future of cars looks different today. Today it's all about self-driving cars. Fuel efficiency isn't as sexy as autopilot mode. According to current thinking, the main design flaw of cars, the obstacle to progress, isn't the gasoline engine, it's the human behind the wheel. The big dream is to remove the human. Google is working on this in a major way. BMW. Audi. Ford. Uber, the ride-hailing company. *Apple* is investing in driverless vehicle technology. And Elon Musk and Tesla Motors of course. Tesla delivered 76,230 electric cars last year and is already producing batteries at a "gigafactory" in the Nevada desert to power its next-generation vehicles. Seven months after the company enabled the use of autopilot mode across its fleet, a forty-year-old Tesla owner died in the first fatal crash of any driver to use that feature. He was watching a Harry Potter video on the car's screen and slammed into a tractor trailer.

More than 30,000 people already die each year in motor vehicle accidents in the United States, and if self-driving vehicles are perfected and become mainstream they will save many lives. They will also destroy good jobs in the trucking, taxi, and ride-hailing industries. Uber likes the technology because it cuts the company's labor costs. As I write this, Uber is running pilot tests of autonomous vehicles in Pittsburgh, Pennsylvania and Tempe, Arizona. Sixteen driverless Volvo SUVs are now available to Uber users in Tempe, picking up riders and carrying them to their destinations.

Please understand: Almost no one a decade ago saw this stuff coming so quickly. General Motors filed for bankruptcy in 2009. That was kind of a bummer. The first mass-produced electric car from a major U.S. automaker, the Chevy Volt, was barely out of the concept phase and not yet available for sale. Apple was still selling the iPod Classic. Steve Jobs was still alive. Google's big recent product launch was Chrome, a web browser. And Tesla Motors—well, Tesla was just another precarious Silicon Valley start-up. At the end of 2008, the company veered close to bankruptcy and, not long after that, Elon Musk admitted in a divorce filing that he was personally broke.

Here's what I'm trying to say: It wasn't crazy, in the late 2000s, for garage hackers and entrepreneurs and students and various passionate weirdos to think they could design a better car than the big companies, a car that could change the world. And it wasn't crazy for the inventors to think they had a shot at getting it mass-produced, thanks to the Automotive X Prize, which would focus the attention of the world on their designs, their ideas, and their dreams. It was a strange moment in U.S. history. The insiders—the large American automotive companies—were humbled and weakened, and a window seemed to be opening for any number of outsiders to take their best shot. It felt important to try. Here was a hard problem that needed a solution, the sooner the better, before the economy got even worse, before rising global temperatures elevated the sea levels and flooded coastal cities. The news about climate change wasn't getting any better in 2007 and it's not getting any better today. The president then was George W. Bush, who said he agreed with the overwhelming scientific consensus that human activity is the cause of climate change. The president today is Donald J. Trump, who has said, falsely, that global warming is a hoax invented by China to harm the American economy. He talks about invading other countries and taking their oil.

For me, reporting and writing this book was an escape, an antidote to a feeling of hopelessness. I saw some people who were excited about building machines that mattered, and I got caught up in that excitement. I started shadowing four different groups of inventors. I watched them work in garages and on race tracks from Philadelphia to Southern California. I was there when the teams' prototype cars broke, or overheated, or caught fire at speed. And I was there when the teams improvised fixes and upgrades in brutal conditions and coaxed these machines into doing things they weren't supposed to do—things that, years later, looking back, still seem heroic to me, and hopeful, and beautiful.

Jason Fagone
Philadelphia, 2017

part one

shakedown

The mentality of the man who
does things in the world is agile,
light, and strong.

The most beautiful things in the
world are those from which all
excess weight has been
eliminated.

—Henry Ford,
My Life and Work

1

Kevin and Jen

We speed west into a bank of mottled clouds and blue-black sky and not much else. We drip with sweat. The launch site is two miles back and counting: the barn where the light is on and the guys are waiting to hear if everything's okay. We try to talk to one another, we shout and scream, but our words get shredded by motor noise, by a sound like a blender gone berserk, whirring and rising to a blastoff pitch.

There are four of us in here: me, a woman named Jen Danzinger, and two men. The men are in front, firing questions at each other in quick bursts. I'm in back with Jen. She braces a hand against the door. Headlights blaze by in the oncoming lane. I wonder how much the other driver noticed—maybe just a silvery bulge, maybe our vehicle's whole weird shape. Elongated fuselage, tapering tail. Does he have any idea what the hell he just passed?

The road beneath us marks the boundary between two Central

Illinois farming counties. We're 220 miles from Chicago, 100 miles from Champaign, 80 miles from St. Louis, and only a couple miles from two major U.S. roads. To the northwest is U.S. Route 66, one of the first American highways, built to let drivers bypass little farming towns. Around here it happens to be the frontage road, or access road, that runs parallel to Interstate 55. If we press our faces to the heat-fogged, northwest-facing windows and look beyond the frontage road, we can see headlights torquing through the heavy blueness on I-55—another, bigger highway built in the seventies so that no driver would ever have to slow down in Divernon, Illinois.

Divernon. An old coal town. The seam tapped out around 1920. The mine closed, the railroad went out of business, and citizens drained away to other, more prosperous places. By the nineties, so many people had left that it no longer made sense for the town to maintain a high school. In a final humiliation, Divernon began shipping its children to a neighboring district seven miles west.

The market crash of 2008 caused barely a ripple here because the pond had been drying out for decades. The Obama stimulus bill of 2009 managed to skip across what was left of that pond like a smooth, flat rock. The only thing resembling economic stimulus in this region had come and gone eight years before, in 2002, when the gruff old man who owned the hardware/beer/auto-parts store in nearby Springfield, the capital of Illinois, only fifteen miles north of Divernon, hung a CLOSED sign in the window, walked back into the dusty recesses of his shop, and shot himself in the head. When police removed his body and explored the building, they found a stash of strange treasures: a live hand grenade, a loaded 1955 Smith & Wesson .32 pistol still in its original box, and, incredibly, a secret second floor filled with vintage cars and motorcycles, including a 1912 Harley-Davidson with an intact carbide headlight. A rumor went around town that Jay Leno, a voracious car collector, had heard about the find and made inquiries. Another rumor had it that David Letterman was going to pay top dollar for the Harley just so he could piss Leno off by destroying it on his show. When the vehicles were eventually put up for auction, two thousand people registered to bid, and an equal number showed up just to watch. Two men from Decatur ended up pooling their money to buy the Harley for $80,000.

Our driver, Kevin Smith, a chemical engineer, suddenly throttles down. The vehicle judders to a halt.

"Regen's kickin' in," Jen says.

"Regen" is short for regenerative brakes. In an electric vehicle like this one, the regen system traps the energy normally wasted during braking and pumps it back into the battery pack. The sound it makes is distinctive: not a screech but sort of a *whrr-whrr-whrr-RUH*, like the sound of Pac-Man getting eaten by a ghost.

A smell now fills the cabin, plasticky and coppery, and I notice Jen making a face. She isn't an engineer like the men in front. She's a graphic designer, a programmer, and a onetime creative writing major. Over the past four months, whenever something about this vehicle has confused me, I've gone to Jen for an explanation, because she's sharp and she's funny and she speaks a language I understand. Her role tonight is to pay attention to sounds and smells: every pebble beneath the wheels, every vibration, the smell of 800 pounds of batteries cooking away, inches from her right thigh. This is a test drive. "Diagnosing problems on cars can involve all senses," she'll tell me later, "but I caution against using taste." Jen says that smells are particularly revealing. Her descriptions of automotive smells have a literary quality. Burning oil "is an easy one: the bluish-black cloud, and a gagging pollution at the back of the tongue." Hot brake pads are "like hot metal and pencil tips that threaten to clog your nose hairs." Leaking antifreeze is sticky-sweet, like burning maple syrup.

Hot batteries are a new one, though. She'll tell me later that they smell like Barbie swallowing a roll of pennies and going to a tanning salon.

Now that we're stopped, Jen and I flick the latches at the bottom of our doors. They open vertically, not horizontally: gullwing doors, usually found in vehicles of outrageous luxury, like high-end sports cars. Also the DeLorean from *Back to the Future*. I forget to hold on to my door's leather handle as it rises, as I've been instructed to do, and it snaps up too quickly, causing a loud bang.

Kevin Smith raises his eyebrows. He steps out of his seat and examines the scratch my door made on the roof.

Kevin tells us something is wrong with the car. A few minutes ago, it "didn't want to shift" and "made a bad noise and a bad smell." He confers with the freckled, red-bearded man in the shotgun seat, Nate Knappenburger, who's hunched over a laptop.

With the doors open, we can get our bearings. There's a field of corn to our left and a field of soybeans to our right. A summer breeze ruffles the crops. It's June 2010. Fireflies dimple the corn with light. The air smells like pollen. The sky is pierced with stars.

"Feels roomy all of a sudden?" Kevin says to Jen, his wife.

"Big pimpin," Jen replies.

KEVIN AND Jen have known each other since they were kids. They went to the same elementary school, lived in the same town: Park Forest, a middle-class suburb of Chicago. Kevin's father managed a grocery store there. Jen's parents were divorced. She lived with her mother, who worked as a dental assistant. Their houses were a quick bike ride away, which is how, when they were nine, they first got to be friends, despite having nearly opposite personalities. Kevin was engaging and pointy-eared and never stopped talking. Jen was slight and private, with green eyes and birdlike features. But they both sensed they were different from the other kids; instead of sports, they liked movies and fantasy games. At school, during recess, Kevin and Jen liked to sit on a patch of concrete and draw elaborate Dungeons & Dragons mazes. Drawing the mazes was more fun than actually playing the game. They'd see how complex they could make them, putting monsters around every corner, adding traps and hidden passages.

Kevin had never met anyone like Jen. She liked to climb trees and catch crawdads in the creek for the pure muddy thrill of it. She'd once tried to bloody the nose of a little boy in her class for some foul remark, at which point her teacher pulled her off the boy and scolded Jen for unladylike behavior. She was cool. She wasn't a girl—she was *Jen*.

In sixth grade, Kevin handed her a marriage certificate he'd bought from a novelty vending machine at their local arcade. He'd taken the liberty of filling out the blanks.

At first, Jen didn't know how to process Kevin's affection. She kept him at a distance until she was sixteen, when she started learning how to drive her mother's car. One day, after school, Kevin insisted on walking Jen out to her car, and before they said goodbye, she saw Kevin looking at her. There was "something about the way the parking-lot sodium lights illuminated his eyes," she would recall later. "Suddenly I felt an emotional connection and realized I'd wasted a lot of time. He was the one."

They started dating around the time they started working on cars together. Their first project was Jen's 1981 AMC Spirit, a classic American crapbox—one of those small, lumpen hatchbacks that American automakers had started to pump out after the oil shocks of the seventies. Because of its powder-blue coat of paint, Jen had dubbed it the Smurf. It was what she drove to her after-school job at Arby's, and it was what she drove to avoid having to ride the school bus, where bullies had singled her out, teasing her relentlessly about the way she looked: her feathered dome of hair, her denim jacket and Iron Maiden T-shirt, her ripped jeans, her skull earring. The Smurf was freedom, and she'd grown fond of it despite its many quirks, the way you might grow fond of some damaged animal you'd rescued from the side of the road.

Of course, it eventually broke down, during one of the cold-est Chicago winters anyone could remember. In Jen's family, when a car broke, you took it to a professional, but Kevin insisted they could fix it themselves. For years he'd worked alongside his father and his older brother, fixing old Chevys and Ramblers. He went to an auto-parts store and bought a new carburetor. A few nights later, he drove to Jen's house. It was dark and viciously cold, subzero with the windchill. Together, Kevin, Jen, and Jen's mother pushed the dead Smurf from the driveway into the single-car garage. They couldn't get it completely inside, so they hung blankets on either side of the car to cut down on the wind. The cold made the metal tools painful to hold. As Jen's mother kept the kids supplied with mugs of hot chocolate, Kevin showed Jen, patiently, step by step, how to replace the carb. *You do this. Then this. You unscrew this.* She'd always thought there was some magic to working on cars, some secret instinct. Now she realized there was no magic, only screws and bolts, belts and hoses.

After that, Kevin and Jen shared a passion for puzzling out automotive problems. Throughout college, they patched up each other's cars so they could drive to see each other on weekends. Jen attended a small liberal arts school not far from Springfield, while Kevin studied chemical engineering at the University of Illinois at Chicago. He joined the Society of Automotive Engineers club, which built alternative-fuel cars and entered them in competitions with other schools. The club didn't have much money, so it had to be creative. Parts were foraged from Dumpsters. Engines were bizarrely configured. A solar car included a seat stolen from a lec-ture hall. A hybrid Dodge Neon used a Korean-made diesel engine meant for a riding lawn mower.

The Neon almost got Kevin killed before he could earn his degree. In a 1995 competition, he was riding shotgun, at a race-track, when a ten-inch cog came loose inside the engine. Spinning incredibly fast, the large silver cog bounced around the engine compartment, finally shooting out the bottom of the car through the wheel well and snapping one of the car's metal A-arms, part of the front suspension. The Neon went skidding into the grass. Jen was there that day, watching in the stands, and when Kevin's car didn't emerge from behind a tree stand on the racetrack's far straightaway, she ran down the steps and sprinted across the infield grass. When she got to the other side, heart racing, she saw Kevin with this huge grin on his face, like he had just experienced one of the greatest moments of his life.

By now, Jen had graduated from college and was working on her master's degree in English at a school in Springfield. When Kevin graduated in 1996, he got the first job he could find in Springfield so he could be near her. He became a bureaucrat with the Illinois Environmental Protection Agency. He wrote pollution-prevention permits for the state. He calculated the toxic emissions of steel factories and huge agribusiness plants, then produced reports studded with chemical symbols and descriptions of manufacturing processes. "Like most people," he would say later, "I have a job that I hate." But the pay and benefits weren't bad, and by 1997, he and Jen had saved enough to buy their first home, an $80,500 ranch house on the outskirts of Divernon. A farmer had owned it for years, and before that, a coal miner. Kevin loved the house because it came with a barn, a creek, and numerous mature trees; an avid hiker, he'd always dreamed of caring for his own piece of wilderness, far from any city. For her part, Jen thought the house was merely "solid," but she soon grew to enjoy living there, mostly because the setting was so private that she could see every social interaction coming from miles away. If you didn't bother your neighbors, they didn't bother you.

The previous owner had built a pole barn in the back—a common, low-cost type of barn often used to house tractors, consisting of a series of poles sunk into the ground with metal sheeting around them—and pretty soon, Kevin and Jen were working on cars in the barn, side by side. For a time, Kevin hacked away at what he called a "FrankenFiero"—two halves of two Pontiac Fieros welded together down the middle, lengthwise. Meanwhile, Jen tinkered with an old AMC Gremlin. Its original firecracker-red paint job had faded to a sad orange. She appreciated the car's ugliness and its ridiculous, unmarketable name. It was like a monument to confused corporate thinking—the ultimate underdog car. Later, she would pay to enter it in a car show, just so she could see the looks on people's faces as they walked by: '57 Chevy Bel Air, '68 Plymouth Roadrunner, '27 Ford Model T custom roadster . . . *Gremlin*.

As soon as they bought the house, Kevin and Jen worked to pay off their small mortgage as quickly as they could. They'd always been afraid of debt. It was best to live simply and owe nothing. That was freedom.

In the summer of 1998, a local Unitarian minister married them in an outdoor ceremony behind their house, beneath the canopy of a big white oak tree. Their friends stayed for two days, drinking beer and eating barbecue and cake. At that point, Jen started working three part-time jobs on top of her day job as a programmer at a Springfield business. She'd promised Kevin she would pay off $10,000 in credit card debt she'd been carrying for years.

One year became two, two became five. They worked, saved money. Jen switched jobs, going to work for an association of electric cooperatives, member-owned groups whose mission was to bring electricity to rural areas. She developed websites, designed newsletters and logos, and laid out magazine articles for *Illinois Country Living* magazine. Kevin stayed where he was, at the Illinois EPA. Instead of having kids, they adopted two dogs that others had abandoned: Scooter, a Pomeranian mix, and Sala, an Australian shepherd with eating issues. To cut down on heating bills and make themselves more self-sufficient, they installed a geothermal heating system, then paid it off early. By 2007, Kevin and Jen owed nothing on their credit cards or their two cars—a Chrysler Neon and a Ford Ranger pickup. All Kevin owed was a small monthly payment on a Harley-Davidson motorcycle.

Then, that fall, he heard something about a contest.

I FIRST became aware of Kevin and Jen during a phone call with an X Prize staffer in February 2010. I'd asked the Foundation to point me in the direction of a few interesting teams, and a guy there suggested I look into Illuminati Motor Works, which is what Kevin was calling his project. "Battery-powered dreamboat," I wrote down. "Illinois cornfield. White guys."

I live outside of Philadelphia, Pennsylvania. I left a phone message with Kevin in Springfield. He returned my call on his lunch break, from the cab of his pickup truck. He talked faster than I could keep up; a brief call yielded five pages of quotes about batteries, aerodynamics, welder's gloves, stoves, and glass, each strand of thought hopelessly knotted into the next.

I wanted to understand. So in March 2010, I flew from Philly to St. Louis. I rented a car, drove two hours north to Springfield, and slept in a motel.

The next morning, I rose early and headed south on the interstate. After twenty miles, I exited onto Route 66, pulling past grain silos and fields of soybeans in a low, thick fog. In the fields, slow-moving tractors seemed to smoke like freshly doused lumps of coal.

Before long, I approached a rare vertical blip in the landscape—a thick circle of tall trees concealing some kind of compound. I pulled into a gravel driveway and parked beside a small burgundy house with white trim. Off to the left was a creek that led to a barn. Beyond the trees lay cornfields as far as I could see.

A black-haired man stepped over two lethargic dogs to answer the door. He was wearing a hat that read SCOTTY'S QUALITY RECYCLED AUTO PARTS. There was a can of generic cola in his left hand. His gray T-shirt had a welder burn in it the size of a tennis ball. He grinned hugely, showing off a pair of cherubic dimples. Kevin Smith had told me he was thirty-nine, but he probably could have passed for twenty-eight. He had a brown mustache, a soul patch, and a thin crescent of beard on his chin.

"Right now I'm on a furlough from the state," he said, beckoning me into his kitchen. "Illinois is in trouble."

Because the recession had drained the state's budget, Illinois had offered state employees like Kevin the chance to stay home without pay. Some of Kevin's colleagues had taken the furlough to spend more time with their spouses and kids. Kevin took the furlough so he could work on his X Prize car.

"I've spent everything I've made for the last two years on this project," he told me. "Last year's taxes were interesting. I actually spent more money than my taxable income. It's a voluntary furlough, and I'm like, *Woo-hoo! I need the time.*"

At the kitchen table, three men in flannel shirts sipped coffee. They were also on furlough from state jobs.

"This coffee is terrible," one of the men told Kevin. "I'm sorry. It really is."

The men picked up their coffee mugs, walked over the two napping dogs and out the front door, and took a left. They passed a filled, aboveground pool covered with green scum. Beyond the pool was a barn with a heavy maroon door. Kevin opened the door and flicked on a light, revealing a cavernous space littered with tools and auto parts. Shelves groaned with stray gears, bolts, pieces of plywood and metal, and buckets of epoxy. In one corner was a makeshift, scale-model "wind tunnel," a plywood contraption that looked like part of a skateboard ramp. A message was scrawled on the wall in blue chalk: SOMEBODY HAS TO DO SOMETHING. THAT SOMEBODY IS US!

Kevin pointed to a small wood-burning stove: "This is where we forged our steel."

I didn't know what that meant. Steel for the car? I thought he was kidding.

He lifted a sheaf of old newspapers from the dusty concrete floor, placed them in the stove, and lit a match. He fed the mouth of the stove with scrap pine until the pine caught. He crumpled a pizza box into the fire, and the flames licked higher.

Then he stood, walked toward the middle of the shop, and proudly announced, "Here's the car!"

The entire surface of the car's body had the ridged, shriveled, dessicated texture and yellow-brown color of dead skin. And there was a *lot* of surface area. In front, a pair of large, bulbous fenders swooned up and over the wheels, a style reminiscent of luxury cars from the thirties, and the car sloped back from there in a gradual taper of shocking length, stretching on and on, like the trail of a diva's gown. The dead-skin stuff was dried epoxy. Soon the car would have a coat of primer, then a coat of silver paint atop the primer, but for now, Kevin and his buddies were still in the process of gooping epoxy on the car's fiberglass body, which they had built up, layer by layer, entirely by hand, like middle schoolers doing papier-mâché.

Kevin explained that under the fiberglass and epoxy was a steel-tube frame, also built by hand. I didn't know you could build a car's steel frame by hand, but there are lots of things I don't know about cars, and Kevin made the frame seem like not that big a deal. He'd simply gone to the local metal supplier and scrap yard, Mervis Iron, and bought a bunch of steel tubes at anywhere between 30 and 90 cents a foot. The steel was regular old "mild" steel—the kind used to make bridges—and not the more expensive, lighter chromoly steel used to make bicycles. He'd brought the tubes back to the barn and stoked the wood-burning stove until the coals were orange. Wearing a welder's glove, he'd gripped a length of steel tubing and fed one end into the mouth of the stove. Once the end was glowing red, he'd carried it to a workbench where Josh Spradlin, one of his friends and colleagues, had clamped a piece of plywood that formed a curve: perhaps the sharp curve of the car's fender, or the curved, arching rib that would form the top of the car. Kevin wrapped the molten tube around the top of the plywood. Josh held a hammer in each hand, pinning the tube to the plywood with the hammers as Kevin gripped a third hammer and beat down on the tube, forcing it to conform to the curve.

Once the steel tubes had cooled, Kevin welded them into a frame. He told me the car contained several thousand welds. This seemed like a lot, especially if they were all done by hand, and by one dude, instead of by those slick-looking robots you see in car commercials. I couldn't shake the image of a crucial weld in Kevin's car snapping like a chicken bone at 70 miles per hour.

"Doesn't the structural strength of your car depend on your skill as a welder?" I asked.

Kevin puffed himself up and flexed his arms. "Why, yes, it does," he said, grinning. "Thank you!"

"Thomas will be the first one to drive it," Josh said, nodding at Thomas Pasko, a team member who runs a local auto-repair shop.

"My wife just took out a $2 million insurance policy," Thomas said.

"We should all take out $2 million policies," Josh said. "We seat four people in the car."

By now, Jen Danzinger had joined the men in the shop. "My wife's smiling," Kevin said. "Apparently she already bought a policy."

Jen tilted her head thoughtfully.

"I could get a new couch," she said.

Kevin turned serious for a moment and admitted that, yes, he was "paranoid" about something breaking. It wasn't just the welds. Much of the car was made from used parts that Kevin and his friends had scavenged from local junkyards. The rear suspension

Illuminati's car in March 2010.

came from a Dodge Neon. The struts were a combination of Nissan, Miata, and Neon struts. "The steering is Honda," Kevin told me, then frowned, racking his memory for more examples.

Josh jumped in. "The master cylinder is Volkswagen." He was a large man with a long graying beard and a shaved head. His right calf was covered with a tattoo of a hiker in the woods, a colorful scene that was accompanied by a quote from *The Lord of the Rings*: NOT ALL THOSE WHO WANDER ARE LOST.

The windshield was from a Mazda Miata, Josh went on to explain. "Bumpers are '71 Camaro bumpers." He shoved his hands in his pockets. "The front springs . . . Pulsar?"

"No, the *rear* are from the Pulsar," Kevin said. "The fronts are from somewhere else."

"An RX-7, right?" Josh asked.

"I don't know anymore," Kevin said.

"Why the gullwing doors?" I asked.

"'Cause they're cool as hell," Kevin said.

"Easy entrance and egress is the main point," Josh said. "But it looks fuckin' cool."

"It looks fuckin' cool, plus they're really hard to make and somebody said we couldn't do it," Kevin said.

"We've gotten very little local support," Josh said. "Most people just don't believe us."

I asked for some figures. What was the car's range on a single battery charge? What was its MPGe?

"About 500," Kevin said.

Five hundred for *both*.

Five hundred MPGe.

Five hundred miles of range.

"Doesn't that seem wildly optimistic?" I asked.

Aren't you full of shit?

"We thought so, too," Kevin said, nodding. But then he had a friend at work, some guy named George, check his math, and George came up with the same numbers. So Kevin figured he was solid.

I STOPPED asking questions for a while so the team could get work done. This was March. The test drive through the corn and soybeans, the one with me and Jen in the back and Kevin and Nate in the front, wouldn't come until June. Between now and then lay an obstacle known as Shakedown—the first competition stage of the X Prize. It would be a practice round, a chance for the teams to work the bugs out of their cars and get used to the NASCAR track in Michigan where most tests would be performed. "The point of the Shakedown is to try to break your car so it doesn't break later," Kevin said.

In a little more than a month, he and his crew would have to haul the car to Michigan. But right now it was nowhere near ready for Shakedown. The 200-horsepower electric motor—a foot-and-a-half-long cylinder encased partly in iron—was on a table in a side room of the shop. The batteries, dozens of them, rectangular slabs of blue plastic that looked like mutant LEGO blocks, were lying on the floor. The shocks weren't installed yet. I told Kevin that, according to the pictures of other teams' cars I'd seen online, he seemed pretty far behind. "Right," he said—because other teams had chosen to adapt existing cars. "We had to build an entire car and think about all the components that go in it," he went on. "Making all the couplers. Windshield wipers. Hooooooolllly *cow*, windshield wipers are tough! Trying to fit 'em in there and make it waterproof so nothing comes into the car. Just getting the windshield was a big hassle, trying to find something that fit our car. Well, we redesigned the car so the windshield fit. We made it a half inch wider."

Josh disappeared into a side room and came back with what looked like the door of a gym locker. It had a vent at the top. He said it might be useful in building a vent on the side of the car; the electric batteries would get hot, and there had to be a way to let that heat out. "We can spend a weekend trying to build vents," Josh said, "or we can just cut these out, paint 'em."

Kevin examined the locker door. "I was just trying to figure out if they'd work, the size and the style."

"These were probably made in the forties," Josh said.

"The fan was going to go—somewhere," Kevin said. "And we'd have some kind of vent. Now the fan will go where the vent is." He looked satisfied.

Hours passed. Classic rock played from a beat-up radio: Pat Benatar, Neil Young. Josh cut the vents from the gym lockers. The shop smelled like singed metal.

Toward dinnertime, I rode with Jen to get pizza for the team. It turned out to be half an hour's drive through dark terrain, silos parallaxing by. It started to rain. I asked Jen if it was true that

Kevin had spent all of his income for the past two years on the car. She nodded, and the side of her mouth curled up slightly. "Kevin is usually very careful with his money," she said, flicking the windshield wipers to maximum. "So this is an amazing gamble. From the very beginning, we had discussed how this was going to go. We agreed that if we didn't get sponsorship, significant sponsorship, he wasn't going to do the project. And yet, well"—she laughed nervously—"in for a penny, in for a pound."

2

Space Kids

He strapped on his flight helmet and climbed into the cockpit of the cigar-shaped plane. His name was Michael J. Adams. He was about to test an experimental plane called the X-15.

It was 1967. For seven years now, NASA had been using the X-15 to investigate the outer edges of the atmosphere and the first tendrils of space. A "rocket plane," the X-15 was designed to be dropped at an altitude of 14 kilometers from a B52 mother ship called the Balls 8. In a series of test flights, the agency had pushed the rocket plane higher and higher, to 80 kilometers, then 90, then, finally, 100.

At that altitude, a pilot's toes, packed into his pressure suit, went numb from the cold. If he unstrapped his harness, he floated in midair, weightless. He might even look down and experience a sensation of oneness with all creatures living and dead. U.S. pilots are considered astronauts if they reach an altitude of 80 kilometers; for others, it's 100 kilometers, an altitude known as the Karman Line, commonly considered the "edge of space."

That day in 1967, Adams hitched a ride into the sky on the belly of a B52. At an altitude of 45,000 feet, he unhooked from the mother ship above a dry lakebed in Nevada. He dropped to a safe distance and ignited his rocket engine, pointing the nose of the plane almost straight up in the air. He climbed to just above 80 kilometers. Then, slowly, the plane began to drift off its heading, yawing from side to side.

For some reason, possibly because he was experiencing vertigo, a common hazard faced by the rocket-plane pilots, Adams didn't notice the yaw, and neither did the NASA staff on the ground, who lacked a good way to monitor heading. The NASA controller, Pete Knight, told Adams that he was "a little bit high . . . but in real good shape."

Fifteen seconds later, Adams reported that the craft "seems squirrelly." Then, twenty-three seconds later: "I'm in a spin, Pete."

According to the official NASA history of this flight, one reaction on the ground was "disbelief; the feeling that possibly he was overstating the case."

"Let's watch your theta, Mike," Knight said.

"I'm in a spin," Adams said again.

"Say again," Knight said.

"I'm in a spin."

"Say again."

This time there was no response from Adams. He was shooting toward the ground, out of control, upside-down, at a speed of more than Mach 4. Ten minutes and thirty-five seconds after launch, Adams's plane came unglued in the air, shedding debris over an area twice the size of Manhattan.

Not long after the death of Michael Adams, NASA shuttered the X-15 program. And that proved to be the end of the line for the whole idea of a "space plane." For the next thirty-five years, the only humans allowed to break the 100-kilometer barrier were those lucky and gifted enough to earn berths on NASA's Mercury, Gemini, Apollo, and Space Shuttle missions, or on Russian spacecraft. By 2000, fewer than four hundred people—less than one-one-hundred-thousandth of 1 percent of the species—had ever traveled in those ships. And the vast majority were professional astronauts, members of an elite corps of space jocks, as opposed to ordinary human beings.

This struck an American businessman named Peter Diamandis as a pathetic state of affairs. A hardcore optimist with a shock of raven hair and a restless nature, Diamandis had been obsessed with space since he was a boy on Long Island, growing up the son of Greek immigrants: His father was a physician, and his mother ran the family medical office. He watched the Apollo launches on TV. He went to Harvard and earned a medical degree, but he was thinking all the time about space and how to get there without waiting for NASA to swoop down and pick him for a Shuttle mission, which he knew would never happen: He was five-foot-five. For Diamandis to get to space, his flight there would have to be routine, like air travel, and he didn't see why that couldn't be. The main obstacle was cost: $1 billion per Shuttle mission. He had to find a way to reduce that figure drastically.

By 1989, when Diamandis graduated from med school, several private companies had begun launching their own rockets into space, mostly to deliver commercial communications satellites into orbit. Diamandis reasoned that if he could figure out how to get satellites into orbit more affordably, he could speed up the pace of rocket launches and draw new companies into the arena. Free-market competition would drive down the cost of a launch. So he formed his own company, International Microspace, and set out to build a device that could deliver a small, 500- to 800-pound satellite into orbit. But he soon ran into funding problems. "It was a chicken-and-egg situation," he would later tell *Forbes*. "In order for us to attract money, we needed to get contracts." But most big contracts were controlled by the government, "and the government wasn't willing to take any risks."

Besides, satellite payloads weren't sexy. They didn't inspire. Maybe, Diamandis realized, the crucial cargo to launch into space was not mechanical. Maybe it was people: other people like him, space kids who would pay to experience zero gravity, fulfilling their childhood dreams. "Space tourism is the only market with the potential for thousands of launches a year," he told *Popular Science*. "That's how we'll learn to fly routinely and cheaply, just like we do with airliners. And that's what it's going to take to change the paradigm of space."

So he needed a space plane—something small, light, and cheap, more 747 than Shuttle.

Diamandis didn't know how to go about building such a plane until 1994, when a friend happened to give him a copy of *The Spirit of St. Louis*, Charles Lindbergh's 1953 account of his quest to cross the Atlantic in an airplane. Here was Lindbergh, remembering his motivations:

> *There's the Orteig prize of $25,000, for the first man to fly from New York to Paris nonstop—that's more than enough to pay for a plane and all the expenses of the flight. And the plane would still be almost as good as new after I landed in Europe. In fact, a successful trip to Paris wouldn't cost anything at all. It might even end up a profitable venture.*

Wait, Diamandis thought.
The Orteig prize.
Twenty-five thousand dollars.
It might even end up a profitable venture.

The way Diamandis had always heard the story, Lindbergh just sort of flew to Paris for . . . well, for the hell of it. Because he could. Because he was a selfless and brave hero and that's the kind of thing heroes do. But actually Lindbergh flew nonstop from New York to Paris because a man named Raymond Orteig, a mustachioed hotel magnate and aviation buff, had promised to give $25,000 to anyone who could accomplish that particular feat. *Lindbergh was trying to win a prize.*

Diamandis began researching the Orteig prize. The more he learned, the more he came to believe that Orteig was some kind of forgotten genius. Talk about financial incentives! After Orteig announced the $25,000 bounty, nine teams of pilots scurried to win his money. Six men died before Lindbergh's very light craft, the *Spirit of St. Louis,* touched down on an airstrip north of Paris on May 21, 1927, making him one of the most famous men on the planet. Five months after Lindbergh's puddle jump, a new company called Pan Am made its first flight, from Key West, Florida, to Havana, Cuba. Within three years, the number of airports in the U.S. had doubled. Orteig had inspired a man to accomplish what many people had said was impossible, and he'd done it with remarkable efficiency: His $25,000 had induced others to spend more than $400,000, a relatively small price to launch an entirely new industry. It was almost like magic. But it wasn't. It was greed, or competitiveness, or both.

Diamandis started approaching corporations to fund a space prize. He called it, for the moment, the X Prize, imagining that "X" would be replaced with the name of the corporate sponsor. His idea was to offer millions of dollars to any private individual or group (NASA need not apply) that could launch three passengers to an altitude of 100 kilometers, bring them safely back down to Earth, and then do all that again within two weeks. He wasn't thinking about a ship that would go into orbit, like the Space Shuttle. He just wanted the ship to smudge the border between Earth's atmosphere and outer space—to pop up to a height of 100 kilometers and pop back down. These would be "suborbital" flights, like the fifteen-minute bottle-rocket ride that Alan Shepard took in 1961, when he became the first American in space, and subsequent flights of the X-15 rocket plane.

Diamandis was laughed out of many boardrooms. All seemed lost until he met Anousheh Ansari, a wealthy Iranian-American businesswoman who, like Diamandis, had always dreamed of going to space, and who promised to help. In talking with a science fiction–obsessed friend, Robert Weiss, a movie producer whose credits included *The Blues Brothers*, Diamandis hit on a clever way to maximize Ansari's investment. When nonprofits want to raise money for a good cause, sometimes they offer cash prizes—say, $1 million for a hole-in-one at a charity golf tournament—and buy an insurance policy to pay out the prize in the unlikely event that some duffer steps up to the tee and knocks one in.

Diamandis found an insurance company to sell him an unusual hole-in-one policy. (The company was reported to be XL Capital, an insurer in Bermuda.) If any rocketeer satisfied his conditions, the insurer would pay out $10 million. He had until the end of 2004 to find a winner.

Luckily, in early 2004, Diamandis discovered his own Lindbergh: a gangly, muttonchopped aviator named Burt Rutan. The son of a dentist, Rutan had spent his childhood building model planes, and later designed "Voyager," the first plane to circle the globe nonstop on a single tank of fuel. For a long time, he had been thinking about how to bring a vehicle down to Earth from space in the cheapest and safest way, without the requisite heat shields required by traditional space capsules. One night, according to the *New Yorker*, he'd woken up and told his wife, "I know how to configure the spaceship! A shuttlecock."

Badminton birdies. They shoot up at great speed and descend slowly and gently. Rutan's idea was to launch his plane from an airborne mother ship, like the X-15 rocket plane, saving on fuel costs; when it was time for the plane to reenter, the wings and tail section would fold into a high-drag shape, and, like a shuttlecock, it would flutter down.

On October 4, 2004, Rutan, Diamandis, Microsoft co-founder Paul Allen, billionaire entrepreneur Richard Branson, and thousands of other space enthusiasts gathered at an airport in Mojave, California. Allen had pumped $20 million into the development of Rutan's ship. It had two components: a launch vehicle, White Knight, and the ship itself, SpaceShipOne, which looked like a fish attached to White Knight's belly. The two took off, ascended. At 50,000 feet, White Knight dropped the fish, which shot toward space, straight as an arrow. Over a PA system, someone announced that the fish had reached a height of 368,000 feet, equal to 112 kilometers, or 70 miles. When it touched down after a 24-minute flight, the test pilot, Brian Binnie, said, "I wake up every morning and thank God that I live in a country where all this is possible." He scrambled up on top of the ship and waved an American flag.

After the triumph of SpaceShipOne, Branson went on to form Virgin Galactic—the world's first commercial space airline, powered by Rutan's design. He commissioned a fleet of space planes, built a spaceport in the New Mexico desert, and started selling tickets for future rides at $200,000 a pop. (The first flight is scheduled for December 2013.) And he wasn't the only one to start his own space company after watching the Prize. A South African entrepreneur named Elon Musk had been obsessed with Mars ever since he was a kid; he always wondered why NASA was taking so long to send humans there. Concluding that he could get to Mars faster if he started building and testing rockets himself, he formed a company called Space X and did exactly that—and within a few years, Musk won a $1.6 billion contract with NASA to fly twelve supply missions to the International Space Station. What was once a crazy notion—that private industry could succeed where NASA had stopped—soon became conventional wisdom. As Barack Obama would put it in 2008, "We must unleash the genius of private enterprise to secure the United States' leadership in space."

The X Prize Foundation was created explicitly to run the space prize. But after it was over, no one at the Foundation wanted to stop. The idea had worked so perfectly. Couldn't the prize model be applied more broadly? To other needs, other industries? Will Pomerantz, a former Foundation employee who now works for Virgin Galactic, says the thinking was, "Hey, we've got a good thing going here. Why hang up the cleats and call it a day?"

THE STORY of automotive innovation in America in a breath: Everything was possible, and then only a couple things were. All was fast and turbulent and thick with discovery until the industry clotted into a stasis so deep, so enduring, that Americans had a hard time picturing it any other way. But the U.S. auto market was awesomely chaotic once: a weird and colorful splay of small-time tinkerers, strivers, blacksmiths, barkers, bicycle-makers, tricycle-makers, and playboys. Aldo, Altman, Babcock, Beebe, Cannon, Culver, Darby. They made gasoline cars, steam-powered cars, and electric cars. In 1900, when Thomas Edison was working feverishly to make a better battery, 28 percent of all cars in America were electric. The carmakers proved the mettle of their machines by racing them, often on dirt tracks in small towns. The best car won.

In 1901, a no-name engineer born on a farm in Dearborn, Michigan, built such a race car. He was no speed junkie; he dreamed of making "good cars," not fast cars. But the public in those days judged a car by its top speed, so the engineer had to play along. He decided to challenge the most famous racer of his era, an Ohioan named Alexander Winton, to a 10-mile contest on a track in Grosse Pointe. In front of six thousand people, the man from Dearborn wiped the track with him, and earned his first large investment, from a Detroit coal baron. In 1903, Henry Ford beat Winton again, with a different car. A week later, he formed the Ford Motor Company.

He released the Model A in 1903, followed by several other models. By late 1906, in a 12-by-15-foot workshop where he liked to sit in his mother's rocking chair, according to biographer Douglas Brinkley, Henry Ford was directing his engineers to build a car both large and light: "a car for the great multitude," he said, "constructed of the best materials, by the best men to be hired, after the simplest designs that modern engineering can devise." Convinced that "automobiles have been built too heavy in the past," and that metallurgical science had stalled, Ford hired specialists to research a new kind of lightweight steel alloy, vanadium steel, previously found only in Europe. They ended up building the frame of the car out of vanadium steel. That was the Model T, released in 1908: strong, light, inexpensive. In 1913, Ford created the world's first assembly line to feed demand for the car, and within a decade, he was selling almost 2 million Model Ts a year.

At first, he had competition; in 1921, eighty-eight car companies were plying their trade in America. Then came the mass extinction. The first three years of the Great Depression slashed the number of automakers almost in half. Meanwhile, the rich got richer. By the end of the twenties, the Ford Motor Company and two other large and growing companies—General Motors, run by a marketing genius named Alfred P. Sloan, Jr., and Chrysler, launched in 1925 by a colleague of Sloan's—controlled 75 percent of the U.S. car market. Marmon died, Stutz died, Duesenberg died. The "Big Three" swallowed up smaller companies or elbowed them out of business altogether.

As Henry Ford morphed from tinkerer to tycoon, and as the Big Three swelled into empires—building assembly lines, hiring hundreds of thousands of men, buying enough rubber and steel and glass to sustain whole subsidiary industries—it became harder for carmakers to propose something completely different. And if they tried, consumers often punished them. In the thirties, for instance, when Walter Chrysler directed his engineers to make "the best body ever built," they responded with the "Airflow," a teardrop-shaped car with a curved windshield, designed with the aid of a wind tunnel and influenced by zeppelins and the Art Deco movement. According to Vincent Curcio, Chrysler's biographer, the company's engineers believed that the streamlined shape "might not at first appeal to the public, but, if it is fundamentally correct, it will grow in favor and be acceptable," as a Chrysler man put it. Released in 1934, the Airflow was so aerodynamically efficient it was able to set a number of speed records. But the public rejected the car—one writer said it had "a rhinocerine ungainliness which automatically consigned it to the outer darkness of motordom," and consumers complained that it looked like a bathtub—and later models reverted to overpowering the air by brute force.

Henry Ford died in 1947, a bigoted old man raving about utopias and Jews. By the fifties, when postwar real-estate developers were designing and building entire towns from scratch around the needs of the automobile (one was Levittown, New York; another was Park Forest, Illinois, hometown of Kevin Smith and Jen Danzinger), the basic plan of the machine had congealed: four or five passengers, two rows of seats, four doors. A car was a steel rectangle smashed through the air by cheap gasoline and rammed into homes by mass marketing. GM created the appearance of innovation and variety through "planned obsolescence," including the use of annual styling changes like tail fins. Nash merged with Hudson, Packard merged with Studebaker, Crosley vanished. By 1958, there were just five major American car companies left.

And for the next fifty years, automotive was an evolutionary business, not a revolutionary one. Cars got better invisibly—disc brakes, power accessories, safer structures—without seeming to change much outwardly. Congress started regulating the automakers in a major way in the 1960s, forcing them to build safer, cleaner cars by mandating seat belts, crash tests, emissions standards, and fuel-efficiency standards. (Today a car is probably the most highly regulated product you can buy outside of a prescription drug, its every nook and cranny shaped by seemingly endless lines of federal code.) But the automakers fought it all, everything, *hard*. They maximized profits, minimized risk. Fuel-efficiency rules happened to go easier on trucks, so the automakers designed new kinds of vehicles, like the SUV, that qualified as trucks, fattening themselves on the high profit margins. Americans liked big, powerful vehicles, and the automakers came to rely on selling them. And over time, the availability of cheap oil braided consumer behavior and corporate inertia into a knot. It became as hard for the automakers to imagine creating a new kind of car as for Americans to imagine driving one.

Not that no one tried. There were spikes of adventurousness by both automakers and consumers. But every step forward was followed by at least one step back. In 1996, spurred by California regulations that required each of the seven largest carmakers in the U.S. to make 2 percent of their fleets emission-free by 1998, with the number set to rise in following years, General Motors released a fully electric car, the EV-1, a two-seater with a lead-acid battery pack that a driver could plug in to recharge—the first production model of its kind. (A later model of the EV-1 used a more advanced battery chemistry, nickel-metal hydride.) About 1,100 cars made their way to lessees in California and Arizona, and many drivers adored their EV-1s, becoming spirited advocates for an electric future. Eventually, though, in a saga recounted by the hit documentary *Who Killed the Electric Car?*, GM recalled the vehicles, saying that the EV-1 wasn't ready for the wider market. It repossessed the cars from anguished drivers, packed them onto trailers, and shipped them to the Arizona desert, where they were crushed into scrap.

GM could have led the way on electric cars, but it lost its nerve. A few years later, in a similar display of arrogance and timidity, U.S. automakers ignored the Toyota Prius, which Toyota began testing in Japan in 1995 and put on sale in 1997. The Prius married an electric motor and an internal-combustion engine in a "hybrid" design, combining the benefits of both architectures—the efficiency of an electric, the unlimited range of gas. American executives thought no one would buy it, but demand for the Prius eventually surprised even Toyota.

One trend held true across decades, companies, and nations: Cars were getting bigger all the time. Even as engineers improved transmissions and engines, generating real gains in efficiency, designers obliterated those gains with sheer bloat. According to an MIT study, between 1980 and 2004, the average weight of the new cars rose by 12 percent, while horsepower skyrocketed by 80 percent. As transportation expert Daniel Sperling pointed out in his 2009 book, *Two Billion Cars*, when Honda introduced its Accord sedan in 1976, it got 46 MPG and weighed 2,000 pounds. Over the next thirty-two years, the car acquired more than 1,500 pounds of barnacles—a bigger engine, more gadgets—until it weighed 3,567 pounds and got only 29 miles to the gallon.

By the mid-2000s, when the price of oil rose to $70 a barrel and demand for big cars fell, so did the automakers' profits. In 2005, GM and Ford saw their bonds downgraded to junk-bond status. GM lost $10.6 billion that year. "At the beginning of 2007, General Motors, Ford, and Chrysler were like big rowboats careening down the Niagara River," wrote the automotive journalist Paul Ingrassia, "heading toward the falls and the terrifying plunge toward bankruptcy."

And as they fell, what did they leave us? What was the residue of thirty wasted years? *One billion cars.* A billion metal machines coursing through the arteries of the world, defining the range of human motion and thus human potential. A billion cars straining roads, clotting the hearts of cities, clogging landfills when they died. They coughed chemicals that polluted the air and warmed the planet: Every burned gallon of gasoline released 20 pounds of carbon dioxide into an atmosphere increasingly saturated with it. And still more cars were coming, which is why Sperling wrote that "cars are arguably one of the greatest man-made threats to human society"; many of the 85 percent of humans who lacked cars were eager to own them, particularly in India and China, with their rapidly growing middle classes.

It was hard to see how the planet could sustain other car cultures like America's. Still, you could never get rid of cars altogether; there was already too much infrastructure built around them, roads and bridges and cities and car-centric suburbs. Besides, the yearning for personal transportation appeared to be universal: Cars provided not just freedom but economic opportunity. The car problem demanded a car solution.

A complacent industry, a vital human need. For Diamandis, a guy who dreamed of planting orchards on Mars and beyond, was there any riper piece of low-hanging fruit on Earth than the automobile?

HE STARTED by asking for advice. He knew a lot of people. He went to his foundation advisory board, which included Al Gore, Larry Page of Google, former CIA director and green-energy fan James Woolsey, and space entrepreneur Elon Musk, who also owned a start-up that made electric cars, Tesla Motors, named after the Serbian-American electrical engineer Nikola Tesla, Thomas Edison's longtime nemesis. (Tesla Motors had already released a $109,000 electric sports car, the Roadster, and was rumored to be working on an electric sport sedan.) How should an auto prize be structured? What shape should it take? Musk and Gore told Diamandis not to do it at all; an auto prize wouldn't be effective. Woolsey told him he should call a woman named Chelsea Sexton.

A passionate backer of clean vehicles, Sexton had worked on the EV-1 program at GM before the company scrapped it and laid her off. (If you've ever seen *Who Killed the Electric Car?*, you'll remember Sexton; she's the one who does such a thorough job convincing you the EV-1 was a good car you're left with a sense of rage at its demise.) Ever since, she'd been consulting with automakers and energy companies, trying to get more efficient vehicles into production. When Sexton heard that the Foundation was planning an auto prize, it struck her as a great way to build public excitement. "An X Prize is really about capturing hearts and minds more than anything else," she says. "It's that aspirational quality that draws people." Sexton signed on as the Prize's first executive director in 2005.

At the time, Diamandis was thinking in terms of a race. He imagined a cross-country road rally, with stages in nine major cities—an automotive spectacle like the ones of old. Forty, fifty high-gloss pellets of carbon and steel whipping through city streets, men and women and kids lining the courses, clapping and whooping and getting pumped about driving one of those things. The most efficient car would win. And to measure efficiency, the Foundation would use a new yardstick it was developing: not miles per gallon but miles per gallon *equivalent*. MPGe. Before each stage, Prize officials would measure how much electricity or ethanol (or whatever) had gone into the cars, and after, they'd measure how much energy the cars had used. Then they'd convert the energy consumed to the equivalent amount of energy in a gallon of gasoline. MPGe was an easy way to compare apples and oranges and ensure a level playing field.

Sexton thought the target MPGe—the big, splashy number that the Prize would be branded around—should be either 250 or 500 miles per gallon equivalent. A hundred seemed low; hobbyists in garages were already claiming they could get that with their hacked Priuses. Better to aim high. At the same time, she didn't want MPGe to be the sole number that mattered. If the Prize was *only* about efficiency, it wouldn't work. "People run really efficient cars on bike tires and solar cells across the country," Sexton says, "but just because you have efficiency doesn't mean it's something people want in their driveway." To captivate the masses, the cars needed to be efficient and run clean, but they also needed to travel a reasonable distance on a single charge or tank of fuel, and they needed to accelerate and brake and corner and maneuver, and they needed to be safe. Sexton hit on a way to maximize public appeal: a weekly reality TV show, on top of the race. Any viewer could put down a $100 refundable deposit on a car, and the team that got the most deposits while also hitting the targets would win the $10 million—as well as a manufacturing contract.

It was an ambitious plan. But along the way, Sexton's enthusiasm for the Prize eroded. Diamandis was turning out to be a demanding boss; it wasn't unusual to get phone calls from him at midnight or 6 A.M. And they disagreed on a few crucial points, like how to fund the Prize. Diamandis was casting a wide net for sponsors. "He would have taken the Chevron X Prize," Sexton says. "I said, 'We're not gonna have *oil* companies.' He said, 'Oh yeah, Chevron will write a check.' *No.*"

In early 2006, Sexton decided she could have more impact on the industry by working outside of the Foundation. After she left, a team of four people replaced her. One of the new leaders was a guy named John Shore, a former theoretical physicist and entrepreneur who had also worked as a science advisor for a U.S. senator. Instead of a target of 250 or 500 MPGe, the new team settled on 100 MPGe—hard, but doable. "Five hundred would have been impossible," Shore recalls. "And one hundred is a lovely nice round number."

It had taken the Foundation longer than it expected to get to this point, and it had yet to announce the Prize publicly. The delay meant that in January 2007 a large automaker was able to steal some of the Foundation's thunder. That month at the Detroit Auto Show, GM unveiled a new concept car called the Chevrolet Volt. It was GM's first electric vehicle since the EV-1—a plug-in electric with a range of 40 miles on a charge and a small gasoline engine and generator to extend the car's range beyond that. GM officials bragged that the Volt would get 230 miles per gallon in urban driving. "It's a radical departure for GM," company executive Bob Lutz, a cigar-smoking former U.S. Marine fighter pilot and a legend in the business, told journalists at the time. "And it's exciting because it's not a sure thing—there's a possibility of failure here."

Was GM really serious about efficiency, though? Lutz would later be quoted saying he thought global warming was "a total crock of shit."

In April 2007, four months after the Volt's debut, Diamandis released a set of draft rules for the Automotive X Prize at the New York Auto Show. He blitzed TV and radio, hoping that some automotive Burt Rutan out there was listening. "We've been driving the same old internal combustion–engine cars for the last eighty to one hundred years," he told NPR. "And it's time for a breakthrough." He added, "I don't care who you are, where you've gone to school, what you've done before. If you solve this problem, you win the money."

ENTERING WAS easy. All you had to do was submit a brief letter of intent that described your car and a $1,000 check to show you were serious. "We did that, first of all, to screen out the crazies," says Shore. Yet it was sometimes difficult to tell if an applicant was earnest, a charlatan, or insane. One team claimed their car would run on urea. Piss. Shore wanted to ask, *Where's the thing you unscrew to pee in?* But the team was serious. Another team said its car would burn trash as fuel. A car designed by a team in Thailand would be made of foam. Shore was skeptical about that one until the team explained that in the traffic-clogged streets of Thailand, many pedestrians and bicyclists are hit by cars and killed every year, hence the urgent need for a car made of a softer material.

In the video, the team leader put a pig carcass in the foam car, crashed the car, demonstrated that the pig was unharmed, and then roasted it.

Musk's Tesla Motors entered its electric Roadster, and green-energy bloggers immediately tagged Tesla as an early favorite to win. The musician Neil Young submitted a letter of intent, saying he'd hired a team to convert his 20-foot-long 1959 Lincoln Continental to run on alternative fuels. (Young later dropped out, citing time conflicts.) A Texas inventor announced he was entering an electric car that could travel either on roads or on a monorail he'd designed, shaped like a Toblerone chocolate bar. A team of Canadians said they were entering a car they'd sketched on a napkin twenty-three years before. Several inventors advertised radically different kinds of engines (an "external combustion engine," a "ducted blade rotary engine"), and a team from Hong Kong promised its electric car, the Salamander, would "drive from a primitive garage into a dazzling great green future." One team set up a website that was little more than a scanned textbook on electrolytic fuels. Another team, WIKISPEED, bragged it was building its car using the open-source methods of Wikipedia, with members in Seattle and across the country staying in touch via Skype and e-mail. A South Carolina team calling itself Psycho-Active extolled the virtues of its Psycho Foam, whatever that was. A white-haired man from Apopka, Florida, announced his X Prize bid on the front page of his own alternative newspaper, alongside articles he'd collected on home remodeling and the health risks of pesticides.

The Foundation badly wanted at least one major automaker to compete against the start-ups and garage hackers, believing it would boost the Prize's reach and credibility. Around this time, Foundation employees met with Honda, Ford, Chrysler, and General Motors, inviting them to enter their hybrid and electric cars and assuring them that, in the words of Shore, "We're not out to screw you over." According to one Prize official, the Foundation came closest with GM, whose marketing people urged their bosses to enter. But the engineers balked. "At the last minute, the engineering folks said, 'We don't think we could win,' which was kind of ironic, because our analysis said they could," said the Prize official.

It would look bad if a major automaker lost to some obscure start-up, but even just entering a car could harm its brand, according to John Voelcker, who edits Green Car Reports, a website that follows developments in electric vehicles and hybrids. Because Voelcker lives and works in New York City, he says, "I am neither in green utopia groupthink or part of Detroit"; he sees himself as an outsider to those two worlds, a guy who wants the industry to evolve but also realizes that "it's a tough, tough, tough business." He says, "If you're a legitimate automaker, especially if you have a production vehicle coming out, if it gets photographed and publicized running around on a track with a whole bunch of completely bizarre-looking, strange-looking vehicles, that casts it in the light of a science project and not a legitimate automobile that people might want to buy." According to this argument, even if a major automaker (say, Ford) *won* the Prize, it might have a problem. "Because the car itself that they created to win that competition wouldn't look like any Ford today," says Ronald Ahrens, who has written for *Automobile* magazine since 1986. "And it wouldn't look like any Ford they'd want to have on the street anytime soon, either. But there'd be the pressure and clamor from the press: Are you gonna produce this car? And even if they said, well, we want to produce this, well, their dealers are going to thumb their noses at Ford because they sell pickups." The easiest thing for the big automakers was to ignore the X Prize and hope it wouldn't have much of an impact.

By default, then, the Prize became the exclusive domain of the little guy, which set up an interesting dynamic—a race beyond the race. The future of automotive efficiency was now proceeding on two separate but parallel tracks. On one were the big companies, working on their own cars behind their own walls. On the other were the Prize contestants, inviting journalists into their garages, debugging their cars in public. The automakers were like shadow contestants, refusing to participate but destined to be compared with the winners—or maybe it was the other way around. Who would succeed, the inventors or the corporations? In a few years' time, who would unveil the better car?

By early 2009, the Prize's format was in place. It wasn't going to be a cross-country race anymore. That had proved difficult to plan—too complex, too many unknowns. Instead, most events would be held in the same spot, a NASCAR track in rural Michigan, to provide the judges with a controlled environment in which to conduct their tests. The Foundation had also decided to split the contest into three performance divisions, spreading the $10 million among them instead of giving it to just one team. Now half the pot would go to the best "mainstream" car—a car that seated four or more passengers—and the other $5 million would go to two "alternative" cars that seated two passengers each, either side by side or front to back (a "tandem" configuration). The idea was to allow for more creativity in vehicle design while also making the performance targets easier to hit. If multiple vehicles in a division managed to exceed 100 MPGe and passed all other tests, the winner would be decided by a 100-mile race.

In April 2009, the Foundation announced that 111 teams, representing 136 vehicles, had made it past the initial sniff test to become official entrants. That summer, a second winnowing began, in which the Foundation required that all teams write elaborate technical descriptions of their cars and pay an additional $5,000. One focus of this paperwork was safety: The Foundation didn't require teams to crash-test their vehicles, or even to provide computer models of crash tests, but it did ask teams to explain "how the occupant compartment is designed for crashworthiness . . . to provide adequate protection in frontal, side, and rollover crashes," and to demonstrate that their cars "would be likely to comply" with a subset of the voluminous U.S. Federal Motor Vehicle Safety Standards. Prize officials pored over these submissions and tried to separate the cars that actually had a chance from the fever dreams and delusions. By October 2009, officials had tossed out more than half the applications, narrowing the field to forty-three teams and fifty-three vehicles.

The teams that fell away tended to be obscure. One major exception was Tesla, which announced it was withdrawing its Roadster, a car adapted from a Lotus Elise chassis, to focus on producing its long-promised sedan, a car that would be built from scratch. Other contestants speculated that Tesla dropped out because it realized it couldn't win. Most tests had found the Roadster's effective range to be far less than the 244 miles Tesla claimed. The car might not be able to do the 100 continuous miles that the Prize required.

Who was left? A team from Colorado entering a Prius modified to run partly on hydrogen; a team of engineering students and professors that had built a hybrid coupe from the ground up, out of carbon fiber; the manager of an Indiana hardware store, who entered a Chevy Metro he'd converted to run on biodiesel; and Kevin and Jen, who'd built a retro-futuristic electric car from scratch in an old barn.

Of the American teams, the most prominent was Aptera Motors, a Southern California start-up marketing an electric car called the 2e, a three-wheeled missile with a body made of gleaming white aerospace composite. (When you create a stronger or lighter material by combining multiple materials into one, that's a composite; concrete is a composite, but advanced composites use carbon fiber, plastic, or glass, often held in place by a liquid resin that hardens.) The 2e had $24 million in venture capital behind it, including a multimillion-dollar investment from Google. In 2008, before the Prize split into three separate divisions, *Popular Mechanics* called Aptera the early favorite, writing that the company's car was "way ahead of the pack."

There were several European teams: a German group entering

a three-wheeled cheese wedge of an electric car; a British team of racers building an all-electric sedan with a fancy, expensive carbon-fiber monocoque shell; a team in Switzerland that appeared to be entering some kind of covered motorcycle, painted banana yellow.

Perhaps the cockiest team was one that had yet to be mentioned in *Popular Mechanics* or anywhere else—a team with its own unique brand of hubris that was evident in a slick, cocky video it had posted online.

Titled *Philosophy*, the video opened with a man shot in black-and-white, speaking straight to the camera from a workshop. "Good afternoon," he said in a German accent. "My name is Oliver Kuttner with Edison2." Ethereal female vocals could be heard in the background as the man struggled to convey the immensity of an idea. "We entered the competition," Oliver continued, "and we thought, oh, you know, maybe we'll just design a race car and have fun. And it's been a real wake-up experience. What we have found is that even some of the best experts in the world are not looking at the problem holistically." There was a cut to a shot of Oliver zipping through a sun-dappled country lane in what looked like an open-air dune buggy. The buggy zipped past the camera. A man standing on the side of the road turned to the lens and grinned. He had a thick beard and dramatically sloping black eyebrows; he said his name was Ron Mathis.

"The boss might get a bit too enthusiastic," Ron said. He sounded British.

In the closing frames, Oliver pulled up to a Southern mansion straight out of *Gone with the Wind*.

The final shot displayed a link to a website: www.edison2.com. If you went there, you saw a motto: THE TEAM FAVORED BY PHYSICS.

In an engineering competition, that was quite a throw down. The team favored by physics. Were they kidding?

3

The Abyss of Lightness

Oliver Kuttner digs into a pocket of his jeans and plucks out something metallic that he raises up to the light. He places it in my outstretched hand.

"Two-tenths of an ounce," Oliver says.

The object is an inch-long piece of pale aluminum with a hexagonal shaft and a circular ring. A lug nut. A lug nut holds a wheel to a car. Car-wise, you can't get more fundamental than a lug nut. It feels as light as a cough drop.

"Cool, right?" Oliver says, beaming.

He takes a step backward and studies my face for a reaction. Oliver is an imposing figure, six-foot-four and 240 pounds, with large green eyes. In college he rowed on the crew team—the five seat in an eight-man boat, the brute shoveling coal into the furnace. His white-collared shirt says EDISON2 above the heart, and his slightly bulging stomach pulls the shirt taut where it's tucked into his jeans.

We're in a workshop. White walls, no windows. The building

The Very Light Car in progress.

used to be a Dickies jeans factory, years ago, and this is a lightly renovated patch of the old factory floor—a large, high-ceilinged, squarish room. It smells a bit like a scorched metal cooking pot. Layers of noise: blasts of power tools, snippets of conversation, a radio playing quietly underneath. Over on the far wall, next to a trash can full of curlicue-shaped metal shavings, a man in safety goggles is bent over a metal lathe. Its whine overpowers the chatter of the nearby mechanics, who are dressed all in black. I can't really figure out what's going on; it's like there are pieces of three or four different types of companies smushed into this one room. Depending on where I look, I see a hip architecture firm, a race shop, a used-car dealership, or a machine shop. Classic sports-car posters hang on the walls next to 3-D computer drawings the size of architectural blueprints, and not far from the guy at the metal lathe, two men type on laptops at desks. Hydraulic hoses snake up the walls like vines, and sparks from a welder sizzle onto the lacquered pine floor.

Flickering in my peripheral vision are the Edison2 cars them-selves, four of them, spread across the floor in various stages of assembly. Three are still just frames of steel tubing, lacking bod-ies. The fourth, clad in silver-colored fiberglass, looks like a real, drivable car. The car's profile changes drastically according to the angle of view. From the top, it's roughly a diamond. From the side, it's a bird skull with a pointy beak. From the front, it's Darth Vader's pentagonal helmet. The wheels are housed inside pods that jut out from the body. The engines—small, extensively modified motorcycle engines that run on gasoline—are mounted in the back instead of the front.

Oliver turns to a nearby storage rack, where he picks up a length of aluminum pipe—another small, humble car part. He tosses the pipe in the air and catches it with a smack. "It's just stu-pid," he says.

"Stupid light?" I ask.

Oliver nods vigorously. He starts talking about how much money he has spent to make these parts. They're all custom, he says; almost nothing in the car has been bought off the shelf. His engineers designed the parts from scratch and had them machined out of billets—solid blocks of aluminum and steel. "It's the only way you can do it without building casting molds," Oliver says, "and casting molds are quite expensive." In the future, he says, when he goes into production and makes ten thousand of each part at a time, the cost will go down. A $50,000 wheel will become a $200 wheel; a $65 nut will become a 20-cent nut.

The nut is significant, he explains, because it weighs one-tenth to one-third of a typical nut. I gather that Oliver wants me to appreciate the extravagance of how much money he has spent to shave off such a miniscule amount of weight. There aren't a lot of lug nuts on a car—sixteen to thirty-two, depending on its size. His custom lug nuts might reduce the overall weight of the car by a pound or so. But the fact that he would go to the trouble of reinventing the common lug nut just to shave off a measly pound shows how far he's willing to go to prove a concept.

"General Motors and BMW and Ford, they all have light-weighting programs," Oliver says. "In the process, they figure out how to take 3,600-pound Corvettes and make them into 3,400-pound Corvettes, which is a good thing." The major automakers can make cars lighter around the edges. But each of the four cars now under construction in the Edison2 shop will weigh less than *eight hundred* pounds when finished. The lightest car on the market, the Smart Fortwo, weighs 1,800 pounds. Oliver has taken lightness to a new extreme. He says he finds this idea hard to get across to people just learning about the car, because the automakers have "poisoned the water" with their lax commitments to lightness.

"Imagine being in the twenties, and an airplane company says, 'We're going to go to space,'" he says. "And other people say, 'We're *already* going to space, we fly up to six thousand feet.' You have to explain that what you're really doing is actually going to the place where there's no gravity." Oliver adds, "For most people it's very difficult to understand until you hold something like this," and nods to the lug nut in my hand.

EVERY ONCE in a while, you come across the sort of supremely gifted salesman who only needs a few minutes to drag you into his orbit like a rogue sun. Oliver strikes me as one of these guys. He obviously knows how to split the world into people who get it and people who don't and to make you want to stand on the side of the angels. I'm sure he's pressed lug nuts into dozens of hands before mine. (I learn later that he carries around up to twenty lug nuts at once; whenever he leaves for a conference, he runs over to the shelf and reaches into the lug nut bag and stuffs them into his jeans and jacket pockets, forcing the mechanics to scrounge for lug nuts to put on the cars.) But even as I'm thinking this—trying to stay skeptical, trying not to get swept up by the momentum of the pitch—I'm also thinking that the thrust of what he's saying is pretty intuitive. And also kind of refreshingly un-gadgety.

Lightness! A cardinal virtue of human vehicles going back to the covered wagon, the chariot, the dogsled. Others interested in shaping the future of the car like to talk about electric vehicles, hybrids, charging infrastructures, upgrades to the electrical grid, and Google's prototype robotic cars, which are so smart they can drive themselves. But Oliver seems to be after something more fundamental. Use any type of energy you want, he says—gasoline, electricity, hydrogen, compressed natural gas—but make the car light, very light, and you're ahead of the game. Simple physics. The less mass, the less force you need to accelerate it. Isaac Newton's Second Law of Motion.

At first, this is what I think Oliver is getting at with the lug nut: He has made a superlight car, therefore a superefficient car (even though it runs on gasoline, not electricity), and he has done it by making a bunch of custom superlight parts—not just the light lug nut but all these other parts on the racks of the shop. Light hubs and light bearings, light wheels, light shifter knobs, light seat rails. But the story turns out to be more complex than that.

Because if you took all these light parts and put them into a normal car, it wouldn't work. The car wouldn't be strong, wouldn't be safe. Oliver's car can be eight hundred pounds because it has a unique and somewhat alien architecture that allows it to be that light. The big idea, the major piece of new technology, is a custom suspension—the part of the car that controls vertical movement, absorbing bumps and giving you a comfortable ride on potholed roads. In a normal car, the front and rear suspension takes up a lot of space, the various components sprawling out horizontally and vertically. Oliver's engineers have figured out how to confine the vital parts of the suspension to a smaller space. In their car, the mechanism that allows for vertical travel fits entirely inside the wheel, along with a damper and the brake.

This may sound like a minor change, but it's not. The suspension is one of those things in a car that other things are built around, determining the course of the overall design, bounding its potential. It is to a car what a river is to a city. Edison2 yanks up the river, reroutes it, and builds a new city on its banks. In-wheel suspensions have been tried before, in limited ways, but no company has ever used the technology as the starting point for a new kind of car.

Putting the suspension in the wheels eliminates the need for a lot of the structure, the steel you'd normally need to enclose the suspension and to attach it rigidly to the rest of the car. And once you eliminate those structures, the other pieces of the car can be lighter. Imagine you're a three-hundred-pound person who goes on a diet and loses one hundred pounds; you can suddenly get by with a lighter chair, a lighter bed, thinner floors. If it were biologically possible, you could redesign your ankles. The fact that one piece of the car is light means that the next piece can be light, and the next piece. And when Oliver and his team go to assemble the puzzle of the car, fastening lightness to lightness, they end up with something so elemental, so irreducible, that Oliver feels it's explicable only by a sort of Zen koan: "The car is light because the car is light."

He calls it the Very Light Car.

ACTUALLY, THE first prototype of the Very Light Car, assembled in early 2008, turned out to be *too* light. It looked like a dune buggy. It had two seats, a bare, steel-tube frame with no body, and a 1-cylinder Yamaha motorcycle engine. It weighed 367 pounds.

Oliver decided to try a little experiment. He climbed into the dune buggy and steered it onto a public road in Charlottesville, Virginia, where he lives, an hour northeast of the workshop. He hit the highway and opened the throttle, gunning it to eighty, his crown of curly gray hair whipping in the breeze. Oliver used to drive professionally in sports-car endurance races, and over the years he has built several high-end racing prototypes. He would have driven the Very Light Car prototype faster but he was already on probation for having too many speeding tickets.

"We went to the absolute edge," he says, "to the abyss of how light you can make it." It was strictly an exercise: What was the minimum weight of a car that was viable for real roads? Now that Oliver had his answer, he decided to make the car heavier, because a 367-pound car is just as dangerous as a motorcycle. So Oliver more than doubled the weight, adding two seats and increasing the car's safety, which is the main challenge in making a light car. How do you take out weight and still protect passengers in a crash?

In the main room of the workshop, Oliver grabs a printout from a table and moves nimbly through a pair of ten-foot-high double doors. In an outer hallway, the bare-metal frame of a Grand-Am race car is nestled against a wall next to a potted plant. The front of the car is crunched and mangled.

Oliver holds up the printout. It shows a line on a graph. At one coordinate, the line spikes sharply upward. He says it came from the Grand-Am car's black-box recorder, recovered after a crash. During one split second of impact, the driver was subjected to a staggering 80 Gs—a sensation of gravity eighty times stronger than that found on Earth. Fighter pilots black out at a sustained 9 Gs. "The guy unbuckled and walked away," Oliver says. "These race cars are not fragile cars. They are purpose-built machines of war."

Oliver says that, in racing, cars have to be light so they can win, but they also have to be strong so drivers can survive high-speed crashes. Think about Indy car crashes you've seen on TV; the wheels shear off, and the cars, with their pointy front ends, bounce off walls and other cars at all angles. Traditional cars are boxes with flat surfaces designed to crumple in a crash. But the Very Light Car should skitter away from many kinds of collisions instead of engaging. "Those wheels are essentially side bumpers," Oliver says. "The car is designed in an accident to behave like a judo fighter, a karate fighter. In most accidents, you will end up deflecting."

He hasn't done a crash test yet to prove his theory—right now he can't afford to destroy a prototype—but he plans to. Lately he's been studying crash-test vehicles from major automakers. To the left of the damaged race car are the front halves of a 2005 Smart, a car manufactured by Daimler AG, which also makes Mercedes-Benzes, and a 2007 Scion XB. Both are totaled, their windshields spidered and cracked, their frames pretzeled. Oliver convinced a nearby crash-testing facility to give him a couple of its wrecks. "You die in a Smart every time you crash head-on," Oliver says. "I don't care what Mercedes says."

THERE'S A plan here—a plan to make all of this real, to get Very Light Cars into driveways and onto roads. I'll come to understand the scope of it over the next several months. It may be crazily ambitious, but it exists, and it goes well beyond the X Prize. That much is clear today in the shop. Oliver wants to win the contest and the money. But he also wants to be a company that endures well after the Prize is over. He wants to build something that lasts. And lately, he says, he's been thinking a lot about how to do that.

An obsessive reader of automotive history, he knows all about the guys who've come before him, the dreamers and strivers who've tried and failed to breach the heavy iron door of the market: Preston Tucker, the master salesman who built fifty-one prototypes of a revolutionary car, the "Tucker Torpedo," in the late 1940s, only to see his young company crushed by a stock-fraud investigation by the Securities and Exchange Commission; Henry J. Kaiser, the famous dam-maker and World War II shipbuilder, who started manufacturing his own cars in 1946 and lasted only nine years, despite all his wealth and power, tallying about 800,000 cars in the end; John DeLorean, the former General Motors engineer who managed to crank out about nine thousand eponymous sports cars at a factory in Northern Ireland in 1981 and 1982 until he ran out of money and tried to raise funds by discussing a cocaine-trafficking deal with a guy who turned out to be an FBI informant. (At trial, DeLorean was cleared on grounds of entrapment.)

These are famous stories. They're why everyone knows it's impossible to start a new car company in America. So Oliver isn't going to try to do that. He doesn't want to be a car company. Instead, he wants to license his technology and let someone else go through the hassle of actually making the cars. What he wants to sell is not a car but the intellectual property required to produce one. And he thinks this knowledge will be valuable because of the way U.S. regulations are changing.

The previous year, in 2009, President Barack Obama struck a historic deal with the major automakers, creating, for the first time, a national tailpipe-emissions standard. By 2017, auto fleets will have to emit, on average, 213 grams of carbon dioxide per mile or less—a target a bit more lenient than the X Prize's 200-grams-per-mile figure. Soon, Obama will drastically increase fuel economy standards as well, requiring all U.S. fleets to average 54.5 MPG by 2025. The new rules boil down to this: If the automakers want to keep selling gas-guzzlers in years to come, they'll also need to sell high-mileage cars. Maybe electric cars like the Chevy Volt and the Nissan Leaf will do the trick, but maybe not. What if the automakers can't make electric cars fast enough, or can't sell enough of them to raise the fleet average above the new regulatory bar? What if they decide they need some other kind of high-mileage ringer?

This is where Oliver comes in: He can help them do it faster.

He can save them time. He can also save them money. An oft-quoted figure is that it takes $1 billion for an automaker to develop a new car "platform"—a basic architecture that can be adapted for different models. That money is for the re-tooling of the factory to make the car, plus the salaries of the people who work on it, the designers and engineers and executives and marketers and consultants, as well as for years of testing hand-built prototypes costing hundreds of thousands of dollars each. But Oliver is already making the prototypes. He has been exploring the abyss of lightness for three years now, mapping the wrong turns and expensive blind alleys, and he will continue to chart new territory this spring and summer, when he runs the Very Light Cars in the X Prize. Not much is known about the physics of a very light car. A few foreign automakers, namely Volkswagen and Toyota, have built and tested ultra-lightweight prototypes in recent years, but their data are private, and they haven't announced plans to bring those cars into production. No U.S. automaker has publicly said it's seriously investigating the approach. When a car gets very light, what happens in a side wind? What happens when a truck passes? "You don't know," Oliver says, "you really don't know. You have to build a prototype and go test it. That costs millions of dollars." Here in a small shop carved out of an old blue-jeans factory, Oliver is spending the money that the big companies aren't willing to spend. "Anybody who decides today that they want to go build this, they have to start with a best guess," he says. "And the best guess is only as good as their best engineer. And they may have a *better* best guess than our first best guess and they may have a *worse* best guess than our first best guess, but we're *past* the best guess."

It's worth it for Oliver to spend the money, he says, because the payoff is potentially so huge. He thinks there's room not only for a new kind of car but for a whole new *line* of cars. Just as there is a range of SUVs and compacts and hybrids, there will someday be a range of Very Light Cars, he says. Your "normal" cars—that's the word he uses to describe every car now on the road—will weigh from 1,800 to 7,000 pounds. Your Very Light Cars will weigh from 600 pounds for a spartan model to 1,100 pounds for a decked-out six-seater. The normal cars will get 12 to 45 miles per gallon, and the Very Light Cars will get 70 to 130 miles per gallon. Oliver thinks the Very Light Car can cut a permanent notch in the market.

Which is optimistic of him, to say the least, but not absurd, not crazy, because certain cars have done this in the past. Starting in 2000, the Prius hybrid proved to reluctant automakers that there was a demand for green vehicles; today 4 percent of all cars sold in the U.S. are hybrids. Four decades earlier, in 1959, the British Motor Company released the Mini, which freed up space for passengers by mounting the engine sideways; most of today's small front-wheel-drive cars are shot through with the Mini's DNA. It was invented by a Greek-British designer named Alec Issigonis, who had once built his own lightweight race car by hand, which he called the Lightweight Special. (Later in his career, Issigonis griped about the bureaucracies at car companies: "A camel is a horse designed by committee.")

And of course there was the original Volkswagen, which eventually became known as the Beetle. Oliver likes to talk about the Beetle. Designed in the 1930s by the German engineer Ferdinand Porsche (yes, *that* Porsche) as a car for the European masses, and later championed by Adolf Hitler, the Beetle was defiantly minimalist, with a squat shape and a small engine in the rear, not the front. According to Beetle historian Andrea Hiott, at the end of World War II, when Allied forces occupied the German town of Wolfsburg, where Beetles were made, the Allies decided to let the factory continue to operate, keeping the company alive. By the late fifties, Volkswagen was aggressively marketing the Beetle in America, encouraging alienated twenty- and thirty-somethings to "Think Small." And it worked. Stylistically, the car was like nothing else on the road—a flagrant departure from the past. Yet Americans bought Beetles, in droves, because the car met a need: It was fun, it was cheap, it was well built, and it didn't use much gas. Owners came to love the Beetle *because* it was weird-looking, not in spite of it, and as Beetles rode off dealer lots across the land, merging with the general traffic, a mere engineering solution turned into a cultural icon.

Oliver says he wants to build a "Volkswagen for the twenty-first century": a car with less stuff for a world running out of stuff. A car for a warming, unequal world where the rich get richer and the rest of us scrape by. Cars are essentially priced by the pound. You're paying for the raw materials, the steel and plastic. An eight-hundred-pound car might retail for $20,000, max, in America, and he can make an even more inexpensive model for India and China, he says. If you heard about a car that was safe, fun to drive, ran on gasoline, got 100 miles per gallon, and cost $20,000, wouldn't you want one?

OVER LUNCH at a local Thai restaurant, Oliver holds court, keeping the men of Edison2 entertained with tales of past business capers. He's best known in Virginia as a real-estate developer, not a carmaker. Oliver owns more than fifty properties in Charlottesville and Lynchburg. That's how he makes his living—by collecting rent. But due to the recession, he says, "We're not making money these days."

Oliver makes real estate sound less like a career than a series of lucky accidents and sudden passions. "Part of my real-estate empire is based on asbestos," he says, putting air quotes around *empire*. "I bought the National Linen building. I removed the asbestos *personally*. It took me six months. That was one of my on-purpose real-estate deals, ha ha ha."

The guys from the shop are all here, ten of them, including Ron Mathis, fifty, the lead engineer, and Brad Jaeger, twenty-four, an engineer and driver who's raced Daytona prototypes in Grand-Am endurance road races.

"I tried to buy the USS *United States*," Oliver says. He explains that the ship was an old decomissioned luxury liner rusting in a dock in Philadelphia. Years back, he got it in his head that he would tow the ship to New York, anchor it on a pier, and turn it into luxury apartments. "It's a *big building*," he says. "So that was my thinking. I'm a nut. Three hundred thousand square feet. I thought I could really do it. I would get someone big to come in, like Trump. At that point, you know, I would just be a 2 percent owner, but who cares?"

"What's interesting," says Brad, steering the conversation back toward the Very Light Car, "is that our car has about the same aerodynamic drag, or less, as a motorcycle." Aerodynamic drag is a function of both the sleekness of a vehicle's shape and the "frontal area," the portion that meets the air. Even though the Very Light Car has more frontal area than a motorcycle, its shape is far sleeker—think of how a motorcycle rider sits upright, offending the air—which makes up for its size.

Oliver recoils in his seat. "Doesn't the Very Light Car have less drag than a motocycle?" he asks.

"The same," says Brad.

"No, it's less."

"Are you sure?" asks Brad. "In the tests I did—"

Oliver interrupts: "With a very high degree of certainty—I would put one hundred dollars on it—we have a lower aerodynamic drag than the motorcycle."

"I can show you the numbers when we get back," Brad tells him.

"I don't want to see your numbers," Oliver says, swiping his paw at the air, and all the men at the table burst into laughter. "You are too late for me."

For the rest of the afternoon, back at the shop, Oliver keeps pitching me on the Edison2 vision. The stakes grow ever higher: One minute he's discussing the price of copper, and then he moves on to windmills, and now he's sharing his thoughts about what it means that so many people in China ride bicycles. The connecting thread seems perfectly visible to Oliver, if not to me. It's as if he's been thinking big thoughts for years with nowhere to put them, and now that a journalist is here, pointing a tape recorder at his chin, he wants to upload everything all at once. "With honest information we can come up with solutions," he tells me, cheeks flushed, picking up speed, "and what we would like most is to position ourselves as the people who have spent more hours thinking about this than anyone else on the planet, and I think we have that honor today and as such we want to position ourselves where someone else who recognizes the virtue of what we're doing says: I want these guys on my side!"

Oliver pinballs from station to station, checking in with his mechanics. "What you are seeing here is a day when fifteen different things have fallen into our lap," he says. Legal paperwork has to go out in the mail, because Edison2 is trying to patent several aspects of its Very Light Car technology. There's also a crucial X Prize deadline to hit. Nineteen days from now, Edison2 has to demonstrate that each of its cars has been driven at least 100 miles, or it can't compete. The first on-track phase, Shakedown, begins next month. The odometers on three of Oliver's cars currently register at zero.

Oliver approaches Peter "P.K." Kaczmar, the British crew chief and lead mechanic. A little while ago, I heard him singing along with the radio. P.K. is busy putting brakes on one of the cars—a brake on each wheel, four brakes in all. Oliver figures that two brakes, not four, will suffice for racking up the 100 required miles. He wants to get the car driving more quickly, to be sure he'll make the deadline. P.K. responds that he'd rather not cut that corner. He wants to do the brakes the proper way the first time so that he doesn't have to redo them later. Oliver blurts out, "I don't give a flying fuck how you do it. The reality is that it drives. The requirement is not that it drives *well*. It just has to drive."

A young mechanic, Bobby Mouzayck, whose right biceps is inked with a skull that has exhaust coming out of its cheekbones and throttle parts for eyes, assures his boss they'll get it done.

Oliver walks away, mumbling. "Sometimes you can get so *paranoid* about this shit," he says. He plants himself in the center of the shop floor and announces, to no one in particular, that it doesn't matter if there are only two brakes on the car, because "this car will stop on a shorter distance on two brakes than most cars will stop on four brakes." He brightens. "This car is the dog in the room," he says in his idiosyncratic English, meaning that it's the alpha dog. He continues, "This car outperforms my Volkswagen Jetta in every way. It just *does*."

When Oliver speaks, he tends to monologue, and now he launches into a long one. The topic is risk, which he talks about as if it scares and thrills him at the same time. He says his wife is sick of his long nights in the shop, away from their kids. He says he's taking the biggest gamble of his life, and he hears about it every night when he goes home. He has already spent $2 million to develop the Very Light Car. The money has come from private investors and his own pocket.

"I have to say, I'm very reckless with this, financially," Oliver says. "For a while, I didn't have the funding lined up, but I was pushing ahead anyway, having faith that I would find the funding. And this is an all-or-nothing deal. You get nothing back if you design half a car. So it's risky. And even if we win the X Prize, it's not a given that people will give us money to take it further. From my point of view it's truly a once-in-a-lifetime opportunity. It's probably the most meaningful thing that I could do."

Toward the end of the afternoon, seven elderly men arrive at the workshop—members of an alternative-vehicle club at a Lynchburg senior center. They've read about the Very Light Car in the local newspaper and want to see it in person. This kind of thing happens a lot. People walk in off the street.

Oliver greets the men warmly. He tells them they're from "the best generation"—the last American generation to have invented great things. He says, "We feel we've discovered a whole new market segment of what a car could be." He pulls out a lug nut. "Who wants to hear about the very light lug nut from the big nut himself?"

Then, after a few words about the virtues of lightness, Oliver plants himself next to the shiny nose cone of the silver car and glances, almost sleepily, off to the side—a bit of casual bravado. This prototype is completely functional. It seems to have all of a real car's weight and heft: four wheels, seats, a steering column. Oliver stretches out his right hand. He makes a thumbs-up gesture. Ever so lightly, as the men watch spellbound, he touches the silver skin. He pushes the car across the floor with only his thumb.

4

Ride or Die

n Hall D of the Philadelphia Convention Center, a crowd of suburbanites picks its way through a maze of freshly waxed sports cars. Toward the back of the hall, they murmur as they file past a black coupe with no logo. A few speculate aloud: Ferrari? Corvette? A man wearing a gold chain and hair gel stops to get a better look.

There's no shortage of eye candy in this part of the Philadelphia Auto Show. A few yards away is a Tesla Motors booth featuring two Roadsters, and directly behind the black car is a tent full of blond Pirelli girls with red leather tank tops unzipped to show cleavage. But something about this car has caught the man's attention.

A teenager walks up to him. He's wearing a black Phillies cap turned slightly to the side and jeans that billow around his wire-thin legs. He starts talking very fast.

"We have a Volkswagen engine attached to an Audi transmission in the back. We have an Azure Dynamics electric motor in the front. Now, the biodiesel engine will get between sixty and seventy miles per gallon, the electric will get between forty and fifty. . . . Okay, so, on your typical day, you travel, what, ten, twenty miles, if that? You can travel up to forty miles on electrical power *alone*. That means without turning your gas engine on. So you're doing absolutely no pollution. Do all your business: get in, park it, plug it, get done. And if you wanna get on the highway, turn that electric engine off, turn that biodiesel on."

The man with the gold chain is accompanied by his young son and his wife. He tells the teenager that he owns a construction business in South Jersey. The wife is wearing a leopard-print coat and hoop earings so large they almost scrape her clavicles. The son, wide-eyed, says to the teenager, "You wouldn't have to go to a gas station."

"You wouldn't have to *look* at a gas station."

The teenager's name is Jacques Wells. He explains that he's a high-school student in West Philly. He's here today on behalf of an after-school club at his school: the West Philly Hybrid X Team. The team has entered two cars in the X Prize, he says, and the black coupe is one of them. A poster on a nearby easel says CAN HIGH-SCHOOL STUDENTS BUILD A CAR BETTER THAN DETROIT?

"We actually do have another hybrid supercar," Jacques explains. "It was built a few years ago, a little before my time. It was our first pride and joy. It was called the K-1 Attack. This is the next generation. Now we have two hybrid sports cars under our belt, both more efficient than any production car you can buy today."

The woman says, "It's amazing."

"You're gonna make a lot of money," the man says.

"The initial question was, what should a hybrid car look like?" Jacques says. "And the question was asked at a high school because high-school students think outside the box. I mean, if it was left up to Ford or General Motors or Honda—look at the Prius. Perfect example. Who wants to drive a Prius?"

"It's ugly," the man says.

"Thank you so much," Jacques says, detonating a broad smile. "I thought I was the only one who thought that."

UNTIL HE was fourteen, Jacques thought being a mechanic was one of the worst jobs a person could have. He felt he was meant for something better. He'd grown up in foster care, moving constantly, from neighborhood to neighborhood and school to school, but his grades had never changed; in some of the most dysfunctional, violent public school buildings in America, Jacques earned A's. His favorite subject was math. "Every mechanic that I knew was working at a private shop, and always looked dirty and greasy," he explains. "Always had grease stains on their elbows, and ashy knuckles, and I was like, that will never be me."

But his older brother didn't feel the same way. He had attended one of the last surviving public vocational high schools in city, the West Philadelphia High School Academy of Automotive and Mechanical Engineering, located on a side street in a blighted part of the city, across from a trash bin that often spilled colorful garbage onto the sidewalk. Near the school's front door, spray-painted auto parts were glued to the wall as decorations.

In 2006, Jacques followed his brother's footsteps and enrolled in the Academy. Up on the school's second floor, he took all the normal subjects, math and English and history, and down in a classroom next to the Academy's garage, he took shop. Freshman year, his shop teacher was a guy named Ron Preiss, who used to run an auto-service department at a Philly garage. Preiss told Jacques that if he learned how to fix cars, he could save a lot of money over the years, and he'd always have a marketable skill; cars weren't going anywhere. That sounded pretty good to Jacques. Preiss taught the basics of brakes, suspensions, and alignment, and Jacques found, to his surprise, that working on cars required not just brawn but brains. He got a sense of the complexity of modern cars—by far the most intricate machines most people own. Even normal, plain-vanilla, gas-burning cars are full of hidden computer chips, working in concert. According to the engineering magazine *IEEE Spectrum*, it takes 20 million lines of computer code just to control the radio and navigation system in an S-Class Mercedes-Benz; for comparison's sake, all the flight systems in an F-35 fighter jet only require about 5.7 million lines of code. In a typical car, systems both major and minor—windows, doors, brakes, steering, emissions—are slaves to as many as one hundred chips that engineers have stashed away in every part of the vehicle.

Electric and hybrid cars require additional levels of computing firepower, although, in many ways, an electric car is a simpler beast than a gas-burning car—no oil lines, no muffler, no spark plug, no tailpipe. A battery pack powers an electric motor, which is basically an electromagnet spinning around and around at high speed. Its motion is transmitted to a shaft, and then to the wheels. When the battery pack is out of juice, you "refuel" the car by plugging it into a wall outlet or a public charging station.

The good news about electric cars is that modern electric motors are 90 percent efficient: For every ten units of energy provided to the motor by the battery pack, nine units power the car forward. By contrast, the best gasoline engine is only 25 percent efficient. The bad news is the battery pack. Despite decades of struggle in R&D labs all around the world, batteries remain awkward, heavy bricks that limit the range of the car.

That's why hybrids make sense. Gasoline plus electricity. The gas engine extends the range of the car beyond what's possible with batteries alone, and the electric motor handles some portion of the driving, making the car more efficient. (In some hybrid designs, the battery pack is small enough to be recharged by the gas engine or the regenerative brakes; in others, the battery pack is larger, and the car needs to be plugged in to recharge the pack.) The strengths of one architecture compensate for the limitations of the other.

The downside is the added complexity and weight of two propulsion technologies instead of one.

At the Academy, Jacques learned his way around not just gasoline cars but electric and hybrid architectures as well. He demonstrated so much natural talent as a mechanic that his teachers suggested he apply for an internship with the City of Philadelphia's Office of Fleet Management, which maintains the city's more than six thousand vehicles: water-department vans, police cruisers, heavy equipment. The hard-boiled union guys at Fleet didn't expect Jacques to know much of anything, but within a couple of months, they trusted him to work on cars by himself, largely unsupervised. "Don't tell anybody," says Jacques's boss at Fleet, "but I'll miss him when he graduates. He's always been a bright spot in my day." He pauses. "I'll never say that shit again."

One day in his junior year, Jacques got to talking with Simon Hauger, his math teacher. Jacques liked Simon; he had a way of relating abstract concepts to the real world. "The way he teaches," Jacques says, "he could teach algebra to a guinea pig." Simon said that Jacques should come by the garage after school one day. Jacques asked why. Simon said, "We work on cars."

THE BELL rings at 3 P.M. Kids burst from their classrooms and leap down the stairs. Most file past the metal detector onto the street, but a few stay behind, making their way into a drafty garage and slapping down their backpacks with purpose. A vaguely fungal smell emanates from racks of motor fluid, tools, and spare parts, and a sign straight out of the fifties reads ALL SKIRTS & SHORTS MUST BE WORN KNEE LENGTH. On a wall, a lime-green banner is emblazoned with a quote from Henry Ford: "THINKING ALWAYS AHEAD, THINKING ALWAYS OF TRYING TO DO MORE, BRINGS A STATE OF MIND IN WHICH NOTHING IS IMPOSSIBLE."

While several kids enter a small, brightly lit classroom to the side of the garage and dig into a cache of snacks, others cluster around Simon next to the GT, the black sports car that Jacques was pitching at the Philadelphia Auto Show. Simon says he wants to work on the GT's turbocharger.

"What's a turbo?" he asks Diamond Gibson, seventeen, a native of Liberia.

"A turbo, it uses air," Diamond says.

"It's like a fan or something," says Azeem Hill, a thoughtful junior with freckles and thick glasses.

"If I used the fan that I put in the window in the summer, my box fan, and blew air in, would that push enough air?" Simon asks. "No. That only sucks so much air."

A turbo, he continues, is basically an air compressor—a tool that converts energy into quick bursts of air. The point is to increase the engine's efficiency by allowing it to squeeze more air into each piston. To burn a certain amount of fuel requires a certain amount of oxygen, so the more air you jam in there, the more fuel you can squirt in. "Now here's the hard part," Simon says. "You need to pay attention. Any time you compress anything, what happens to its temperature?"

"It rises," Diamond says.

"Right. It gets hotter. And when things get hot, like in a hot air balloon, what do they want to do?"

"Expand," says Azeem.

"Expand. So our goal for a turbo is to pump air in. So you're compressing air. It's getting hot. You're fighting yourself, right? You need something that cools it off."

Sometimes Simon delivers impromptu lectures like this, and the garage transforms into a kind of classroom. But most of the time he stands back and gives the kids space to figure things out for themselves. The students fan out, tightening bolts, drilling holes in metal plates, and checking in every so often with Simon and the other adults who volunteer their time, including the two shop teachers, Preiss and Jerry DiLossi, a former bus mechanic, as well as the team's manager, a remarkable woman named Ann Cohen. Ann used to be the president of AFSCME Local 1637, the union that represents the civilian employees in Philly's police and fire departments. For years she was a figure of genuine power in the city, respected by mayors and congressmen; she's still on a first-name basis with the mayor. Now that Ann is retired, she wields her influence on behalf of the team, coaxing politicians to tour the shop, appealing to journalists to cover it, and finagling free parts from suppliers. "They're hilarious," she says of the kids. "Sometimes they're just horrible. They couldn't tell you how many *wheels* a car had. They don't achieve. But sometimes they're just amazing."

Another regular in the garage is Kathleen Radebaugh, the Academy's lone English teacher. Every day, all 156 of the Academy's students pass through her classroom upstairs, which means that every one of the Hybrid X Team's students is also an English student of hers. Radebaugh says she joined the team partly so she could make sure the team members were doing their homework.

One afternoon at the garage, a bunch of students are examining the GT, including Diamond. Radebaugh looks at him. "Diamond's in my seventh-period class," she tells me. "Seventh-period today was bad."

Diamond, hearing his name, walks over and gives Radebaugh a look like *Who, me?*

"Diamond, who's the main character in the story today? The Ray Bradbury story we read today."

Diamond smiles.

"Come on," Radebaugh says. She exhales and turns to another of her students.

"Who's the main character?"

The student looks at her blankly.

"I'm gonna go home tonight and cry myself to sleep," Radebaugh says.

IN THE beginning, Simon Hauger was just like Radebaugh, a young teacher full of energy and hope. Then he burned out.

It took a little while. His first year working in West Philly was 1994. He taught math at West Philadelphia High, the red-brick building across the street from the Academy. (In the eyes of the Philadelphia School District, the high school and the Academy are pieces of the same bureaucratic entity, but really they're two separate worlds.) That year, a kid was shot in the lunchroom directly beneath Simon's class. Another kid was shot and killed on the sidewalk outside the school. One day, a student returned to school from a juvenile detention facility, walked into Simon's class, and started beating on another kid. Simon was twenty-five and physically fit—spindly and energetic, he used to run track—but he couldn't break up the fight. "I mean, this kid was *big*," Simon remembers. "It's a little unnerving when you grab a student and you try to subdue them and you can't."

Simon's arrival in West Philly closed a sort of personal loop. He'd grown up here, the son of white liberal parents; his father was a social worker at a city hospital, and Simon attended West Philly schools, public and Catholic. He was the kind of shy, withdrawn kid who could put his head down and burn through math problems but found it hard to express himself in words. In seventh grade, he drew a schematic for a perpetual motion machine; it used magnetism and gravity to move a lever up and down. (When Simon found the picture years later, he realized the machine would never work—friction would overcome it—but he still got a kick out of it. "I was like, holy shit! Pretty clever idea.") After high school, he went to Philly's Drexel University and studied electrical engineering. At night, to help pay his tuition, he worked as a valet parker at a fancy Italian restaurant downtown. Lawyers would hand him the keys to their Corvettes and Lotuses and Ferraris, and Simon would drive these cars for a block at a time, just long enough to sense the latent power under the hood, the rumble of a swift machine in low gear, but not long enough to release any of that beautiful kinetic energy.

On a co-op between his junior and senior year, he worked as an engineer at General Electric in the Philly suburbs, designing circuits, but he soon got scared. He saw guys in cubicles who'd been there for forty years, doing the same job, and he didn't want to be like them.

He watched the movie *Stand and Deliver*, about an influential Los Angeles math teacher. Teaching seemed like the right thing. He got his certificate. Then came that first brutal year at West Philly High.

Simon stuck it out for thirteen more years. He was stubborn. But teaching never got any easier. In 1997, Simon moved across the street to the Auto Academy and continued to teach math and physics. He taught it the way it had always been taught—standing at the front of a class, facing rows of kids in desks. He watched kids drop out and get arrested for drug possession. One year he received a letter from a former student who had always been one of his favorites; the kid could sit in his desk and crank through physics problems like other kids waste zombies in video games. It was postmarked from a prison. Prison was boring, the letter went. Could Simon send some physics problems?

The only thing he could feel good about was happening outside of the classroom, after school, down in the garage. In 1998, a couple of Simon's students asked him for help on their science-fair project, an electric go-kart. The go-kart turned out to be so much fun that the following year Simon and the kids picked an even more ambitious task, converting a Jeep to an electric; later, they converted it to a hybrid. Next they turned a Saturn coupe into an electric car. Simon couldn't believe how hard the kids were willing to work.

These were the same students who couldn't sit still in their normal classes and regularly failed standardized tests, but give them a huge dense technical manual and some tools and they'd spend their Saturdays futzing with transmissions and door panels.

In 2002, the team entered the Saturn in the Tour de Sol, America's premier eco-car competition at the time, a multi-day road rally between Washington, D.C., and New York City. West Philly took on forty other teams, including teams from major universities like MIT. The Saturn was so efficient as to be "mathematically impossible," according to a complaint that MIT filed with the Tour de Sol's organizers, but the complaint was rejected, and West Philly notched its first victory. Then the kids started talking about building a different kind of car. Something hot. The automakers' "green" cars, especially the Prius, were so drab. Simon's students coveted speed. They dreamed of doing burnouts in sick machines. They were American teenagers. Couldn't they build an efficient car they'd actually want to drive?

They ordered a sports-car kit on the Internet. Over a period of two years, using the kit's steel frame as a template, and using free online translation tools to make sense of the instructions, which were written in Slovakian, Simon and the kids assembled a custom hybrid, the "K-1 Attack." An electric motor powered the front wheels, and a Volkswagen diesel engine powered the rear wheels; the diesel engine could accept "biodiesel" fuel made from vegetable oil. Sharky and super-low to the ground, and painted bright yellow, the car looked like one of the $200,000 supercars the kids were always gawking at in car magazines, but they'd built it for $15,000. After several years of tweaking, the Attack was getting 60 MPG and accelerating from 0 to 60 in five seconds, and in 2005 and 2006, West Philly won the Tour de Sol once again.

They performed so well in these efficiency contests *because* of their limitations, not in spite of them. West Philly's opponents fetishized gadgetry, installing the latest engine technologies and testing them on the fly. They "über-engineered," in Simon's phrase, and sometimes their cars simply broke. Simon and the kids preferred to use equipment that was a year or two old and therefore more extensively tested. They tended not to be impressed with the latest models.

One year, two men visited the Auto Academy with their shiny electric Tesla Roadsters. They were driving their Teslas across the country to raise awareness about green energy. They'd just come from New York, a trip of about a hundred miles. One of the travelers, an older man, had driven his Roadster conservatively. He still had about 50 miles of range left in his battery. But the other guy, who was younger, only had 3 miles left. Tesla had long claimed that the Roadster's range on a charge was 244 miles, but nobody who knew anything about electric cars believed that figure. The kids asked the younger guy a few simple questions about the numbers on his dash, and it was immediately obvious he had no idea what kind of car he'd bought. "A light switch went off in his mind," Simon recalls. "He had bought a $109,000 sports car that he had to drive at 55 miles per hour if he wanted to go more than 100 miles at a time."

Later, it seemed like the kids would have a chance to inspect another much-talked-about electric car. In early 2010, General Motors called Simon and asked if the company could visit the school. To demonstrate GM's seriousness about green cars—and to generate some good PR—they wanted to send a Chevy Volt chassis to the Academy, along with an engineer to explain its features. Simon thought it would be a great opportunity to see cutting-edge technology up close.

"Listen up, guys," he told the kids at an after-school meeting, "they're bringing the Volt here."

The kids just stared at him.

"I don't even think it registers," Simon continued, almost talking to himself. "Who's somebody like Wyclef?" (The musician

Wyclef Jean had already visited the shop, freestyle rapping with one of the kids.) "Okay, so now Jay-Z's comin'. Big Money. Or Small Money? Young Money?"

Simon and Ann spent a week preparing for the Volt event, inviting students from five automotive academies across the city. On the day of the visit, they ordered twenty pizzas, unfolded hundreds of chairs, and cleared a hundred-foot parking spot outside of the garage for the tractor trailer carrying the Volt. A PR woman from General Motors arrived early and set up a banner that said AN ELECTRIC, RANGE-EXTENDED VEHICLE! As they all waited for the Volt, Ann got a call from a logistics guy at General Motors.

"Can you get me a flatbed?" he asked.

Wait, what?

"I don't think the chassis is gonna make it," he said. The tractor carrying the Volt had broken down in New Jersey.

A few minutes later, Ann got a call from a second GM employee, this one based in Detroit.

"I'm so sorry," the GM man said. "You heard about our problems. I understand the truck's in New Jersey."

"Can't you swap the tractor?" Ann said. "If the tractor's broken down, can't you put a new tractor on it?"

"We're looking at that."

There was a pause.

"Do *you* have a tractor?" asked the GM man.

Ann had to inform the man that while they had a lot of stuff at the Auto Academy, they did not have their own tractor capable of hauling a ten-ton trailer.

When the Volt engineer finally did arrive—a young woman with an intelligent, friendly manner—she spoke to the kids for sixteen minutes about the philosophy of the car, then asked for questions. A junior named Sowande Gay raised his hand. He had long frizzy braids pulled into a ponytail that fanned out behind him like cotton candy.

"Well, with all of this new technology, how much is it gonna cost?" Sowande asked.

"Yes, that's a good question," the engineer replied. "They haven't released the price of the Volt yet, to be honest, but all of the prices that I've heard range from $27,000 to $40,000." (GM later announced the price at $41,000.)

Sowande frowned. The median income in that part of Philly was $30,000, and more than one in four lived in poverty.

"Did y'all lighten it up? Like, is it light as hell?"

The Volt engineer looked down and nodded. "To be honest, that's probably the biggest downfall of this vehicle. We wanted to be first to market with this type of technology, so we took an existing chassis . . . . Ideally, we would start from scratch and light-weight the heck out of it."

Simon liked how these kinds of exchanges reversed the usual polarity of debates about clean energy. The environmental movement had long been led by white elites, but white elites were the ones *least* affected by carbon emissions and other pollutants; they could afford nice houses away from the pollution and congestion of cities, and if sea levels rose because of climate change, they could move to higher ground. His students didn't have that luxury. They had to live in a world that others had created. But there was no rule that said they had to stay silent about it.

In 2008, Simon quit the Philadelphia School District and went to work for a nonprofit that supports innovative programs in urban schools. The nonprofit paid him to run the West Philly Hybrid X Team, as it was now known, full-time. The move gave him some additional space to operate. But he wanted even more. By now, Simon had made friends with a handful of graduate students and professors at the University of Pennsylvania who were interested in transforming urban education. Whenever Simon got together with them, he and the Penn guys always ended up fantasizing about starting their own school—a place where they could take the latest academic insights about project-based learning and apply them in classrooms of their own design.

This was what had excited Simon about the X Prize when he first learned of it and read through the rules. For one thing, it would be structured like a series of races. If Simon and his students knew anything, they knew how to build fast, rugged, versatile hybrid cars, exactly the kind that would hold up well to grueling race conditions. Even better, the cars had to be "production-capable." Each team was required to submit a detailed business plan showing how it could plausibly crank out ten thousand of its cars by 2014. The Prize didn't appear to be an über-engineering competition. The rules demanded a balance between practicality and performance, a trade-off that was built into West Philly's DNA.

The contest promised to give its champions a global stage. If West Philly won the Prize, Simon could attract money and support for a new school. He could finally break away from the District. In an important sense, he and his students would be competing for their freedom.

THE KIDS stand in a rough semicircle, a dozen of them, murmuring prayers in teenage slang. "Jawn," an all-purpose ghetto noun. "So delicious," a description of what they're staring at: a chaotic snarl of wires, pipes, metal plates, zip ties, and Gorilla Tape. At the center is the silvery, V-shaped engine of a Harley-Davidson motorcycle. Everything is fastened to a pair of blaze-orange sawhorses.

"What are the chances this is gonna explode?" Sowande Gay asks Simon.

"None," Simon says. He pauses and runs his hand across an icy thatch of prematurely gray hair. "You got insurance?"

"Aflac," another student chimes in.

"Then maybe *you* should stand over here," Simon says, "because I've got bo bo insurance, ha ha ha." *Bo bo*: crappy, bad.

He turns to me and laughs. "You think this is what General Motors' bench-testing facility looks like?"

The team has decided to enter two hybrids for the Prize. The first is the sports car Jacques was selling at the Auto Show. This thing on sawhorses is the heart of the second car: a sedan based on a 2008 Ford Focus. A while back, Hauger bought a Focus, drove it off the lot to the team's garage, and got the kids to help him remove the engine. I'm looking at what will eventually take its place—a hybrid architecture of original design.

The philosophy of this car begins with the recognition that the hybrids currently on the road aren't as efficient as they could be. The way to make a hybrid more efficient is to increase the amount of work performed by the electric side. But at this point in 2010, most hybrids still don't have the electric motor doing most of the work. The reason is simple. For the electric motor to run the show, the battery pack has to be large, and when a large battery pack is depleted, the driver has to recharge the car by plugging it in.

People have driven cars for a hundred years without having to plug them in. Automakers worry that drivers aren't ready to make that leap. But the X Prize is about leaping. So the Focus is like the Chevy Volt—a plug-in that runs primarily on electricity but also has a gasoline engine to extend the range. The major difference is that the Focus uses a 1.4-liter Harley V-twin as its gas engine, as opposed to the Volt's 1.4-liter engine, which is adapted from a Chevy Cruze engine and attached to an electric generator. The team went with a Harley because it was about the right size (70 horsepower) and because it was made in the U.S., like the Focus itself. If West Philly ends up hitting the X Prize's targets, they will be able to say that theirs is the first American-made car to achieve 100 MPGe.

Once Simon and his colleagues decided to use the Harley, they had to figure out how to make it play nicely with the electric motor and the rest of the car. The Harley is brawnier than the gasoline engines in other hybrids, and Simon had to make sure it wouldn't break anything. He thought of farm equipment. He imagined a guy in a field driving a big-ass tractor, sucking up wheat and corn. Many tractors have clutches that use magnetic fields instead of spinning plates to handle large power loads. Simon found a magnetic clutch from a tractor. After some tinkering, he discovered he could use the tractor clutch to connect both the Harley and the electric motor to the transmission, the part of the car that transfers power from the engine or the electric motor to the wheels.

And there it was: a plug-in hybrid with weird American guts. A Volt on the cheap.

Now he's about to fire up the whole thing for the first time.

Not far from where Simon is standing, shop teacher DiLossi squats on the concrete floor, holding two wires. DiLossi nods slightly at Simon and touches the wires together, feeding current into the electric motor, which will, if everything works properly, start the Harley growling:

chrrr chrr chrr

chrr chrr chrr

POP

A spark, a sound like a gunshot blast, a cloud of acrid smoke that smells like burning tires.

"Whoa, hold on," says DiLossi, falling back onto his butt. "I just got the shit juiced out of me." He pats his shirt pocket to make sure his Marlboro Lights haven't fallen out.

"I smell something burning," says Sowande, grinning.

"It was my hand," DiLossi says. "Why don't you come over here, tough guy?"

"We can have Michael sit on it," says Ann Cohen. Michael Glover, one of the bigger kids in the garage, is an offensive lineman on the school's football team, the West Philadelphia Speedboys. "Me and Michael will sit on it," Ann says.

"I ain't gonna die," Michael says.

Ann asks Michael to find a sandbag to put on top of the Harley, in case it explodes.

chrr chrr chrr

chrr chrr chrr huagggggghhh

Like a preacher warming to his sermon. Then the engine turns over, and the noise blasts out of the garage and into the neighborhood.

The kids dance and shout. "Like a *lion*," one of them says. DiLossi crosses himself. Ann laughs and says this is the loudest Ford Focus in history. Sowande chest-bumps with the wiry Senegalese senior Sekou Kamara, who chose the team motto, "Ride or Die." A gangly kid spins like a top across the garage, saying, "Oh my God, oh my God DAMN." Even Simon, whose job is to worry about all the things that can go wrong, unleashes a giddy whoop. Then he shoots a look at DiLossi, who's revving the Harley with glee.

"Take it easy," Simon says.

"I wanna hear it," DiLossi says.

"Dude, take it easy."

DiLossi revs it anyway.

5

Immigrants

liver Kuttner's father was a poor man. His name was Ferdinand Anton Koch. An actor and writer in Munich, Germany, he made his living by writing richly illustrated books about the art objects and religious rituals of pre-Columbian cultures: the Aztecs, the Inca, and the Maya. He wrote them under the name "Ferdinand Anton." He wasn't a professor of archaeology and he didn't have a university degree or appointment. He had simply declared himself to be an explorer and then explored. He flew to South America and Mexico and hiked to remote shrines where few Westerners had set foot. He wrote more than forty books. "The ancient Maya ceremonial centre of Palenque lies beneath us," Anton wrote in his 1970 opus *Art of the Maya*. "To the pilot, ruins are ruins, but I, who have come here from the Old World, experience the same thrill as others who have thus unexpectedly come face to face with monuments of a civilization that has long since come to an end. Palenque was built for gods who are long dead, but it is immortal; its heart still beats, ruined and yet perfect, uninhabited but full of life."

After each trip, Anton returned to Munich with bags full of masks and statues that he hung on the wall next to original paintings by Oliver's mother, the artist Beatrix Ost. Oliver remembers that when there was a book contract, there was money in the house, and when there was no book, there was no money.

His mother and father divorced when he was six. At eight he went to live for a time with his grandmother, in an apartment building in Bavaria surrounded by farms. In that era, the late 1960s, it was considered undignified for women to drive, so Oliver's grandmother taught him how to drive her old BMW. By age nine, he was driving her into town on errands. When they passed police officers on the road, the officers would stare in shock at this small boy behind the wheel, and Oliver and his grandmother would smile and wave politely and keep driving. Oliver's grandmother, seeing how much he loved to drive, took Oliver to the occasional hill-climb races on nearby Wallberg Mountain, the drivers twisting up the closed-off road one at a time, racing the clock. She also showed Oliver picture books of her next-door neighbor, a famous German race-car driver named Hans Stuck. Oliver loved those old pictures of Stuck in his 1920s-era Austro-Daimler hill-climb car, the state of the art at the time, a cream-colored cigar.

Oliver immigrated to America in 1975, at fourteen. He arrived the old-fashioned way, on a boat. By now, his mother had remarried, to Ludwig Kuttner, an industrialist, and Ludwig had thought that it would be nice for his family to arrive by sea, like families of yore. Ludwig booked passage on the *Queen Elizabeth 2*, a luxury liner with a deck the length of three football fields. Oliver remembers the ship as something beautiful, mysterious, majestic. The family sailed in September, under gray skies, and when Oliver and his two younger brothers, Fabian and Daniel, would pass time by swimming in the ship's wave pool, the waves in the ocean would amplify the waves in the pool, and the water would crash over them.

Oliver spoke almost no English when he arrived in the States. The first time he took the SAT, at his high school in suburban New York, he scored 270. The only person who seemed to believe Oliver had any potential was his shop teacher, who taught him how to take apart an engine and put it back together, stoking his already heightened appreciation of fine machinery; as a kid in Munich, he had spent endless hours in an automotive bookshop, devouring car magazines, and he remembered almost everything he read.

Oliver had always imagined he would follow in the footsteps of his birth father. He would become some kind of academic, maybe a scientist. But he had seen his father struggle; he had seen how Anton's troubles with money limited his freedom to pursue his passion. Oliver's stepfather, Ludwig, who owned a thriving sweater manufacturing company based in New York and the South, didn't have that problem. "By the time I was fifteen, I realized that it takes money to make things happen," Oliver says.

With money he earned by mowing lawns, he bought his first Ferrari at sixteen. European sports cars were underappreciated at that point in America, the province of specialists. Still in high school, Oliver began to prowl New York's junkyards on weekends, looking for the wrecks of rare sports cars that had been overlooked by junkyard owners. His biggest score came in Brooklyn, where a diminutive, elderly Sicilian man—"He was really connected, like, *connected* connected"—owned an apartment building with more than a hundred parking spots in the basement. The Sicilian had claimed the basement for himself and filled it with cars of dubious origin: abandoned cars, cars recovered after being stolen, cars that had been auctioned off for huge parking fines, cars that had been sold to the playboy son of somebody rich, who crashed it then sold it to another guy, who broke the motor and didn't know where to get parts so he put a Chevy V8 in it. By the time cars got to the Sicilian, they were the automotive equivalent of laundered money, their true origins obscured, untraceable. But Oliver knew exactly what they were.

"I bought half his stuff," Oliver recalled. "I took trucks out of there in volume." Oliver later sold some of the Sicilian's cars for hundreds of thousands of dollars.

Oliver didn't like high school, so he worked to finish early, attending summer school and graduating at sixteen. He applied to Harvard, Yale, Princeton, Cornell, and the University of Pennsylvania. Rejected by all five, he eventually settled on Boston University, where he majored in business management, minored in French, and filled his dorm room with disassembled Corvette motors. It was then that he tried out for the crew team, to get a tan and make friends. He turned out to be a natural—rowing, above all sports, rewards blind tenacity and a capacity for physical suffering—but today his teammates remember Oliver less for his rowing prowess than his unorthodox ideas about how to win. He once begged the coach to let him fill the inside of the hollow boat with helium, to make it lighter; another time, he argued that the team should add a "wing-rigger" to generate lift. It would pick the boat a few millimeters off the water, thinning its wake, making it go faster. Oliver literally wanted the boat to *fly*. "It was all crazy stuff," recalls Stefan Bub, one of Oliver's former teammates, now an investment banker. "But when you sat down and thought about it, you know: actually *not* stupid."

By his senior year, Oliver had gone from friendless immigrant to successful athlete and student-union treasurer. After he graduated, his stepfather arranged a job for him at a bank in Munich. Oliver was miserable there—the strict hours, the serious atmosphere—and he returned to America. According to Stefan Bub, he and Oliver "are probably the most American Germans you can find. It's no coincidence that we both chose to live in this country, simply because Germany, it's too suffocating."

In 1983, Oliver received a letter from Ludwig and Beatrix, asking him how he was doing. They had just moved into a grand and unusual house in the countryside outside of Charlottesville, a house with a fascinating lineage. Its builder, James Dinsmore, had been Thomas Jefferson's longtime master carpenter. Born in Northern Ireland around 1771 and naturalized in Philadelphia, a city known for its machine shops and craftsmen, Dinsmore came to Charlottesville in 1798 to live at Jefferson's Monticello estate. He ran the joinery there, and helped Jefferson complete a series of ambitious renovations. (Dinsmore and Jefferson often exchanged letters. While Jefferson's prose is breezy, smitten, spiced with measurements and drawings, Dinsmore is courtly, matter-of-fact, sometimes hilariously dry. "The old woman Junea is dead the rest of the family all well," he wrote to the President in 1802. "I am still engaged in the Dining Room.") Dinsmore laid the cherry and beech parquet floor in Monticello's parlor; he supervised construction of Monticello's famous dome.

The Irishman died in a drowning accident at age fifty-nine, but not before leaving a final monument to Jefferson's dream of America: Estouteville, a grand private home intended for one of the male children of a wealthy planter and Revolutionary War colonel named John Cole II. Estouteville was basically a mini-Monticello— eight thousand square feet, with a Great Hall, porticoes, soaring Tuscan columns, and fine *bucrania*, decorative carvings of ox skulls, made by Dinsmore's hand. According to the state of Virginia's historic landmarks commission, "Few places so well conform to the popular image of an aristocratic Southern homestead. It meets Jefferson's ideal of an architecturally refined seat suitable for the young Republic's landed families."

This was what Ludwig and Beatrix bought. And this was where they invited Oliver to live.

Oliver accepted the invitation. For as long as he could remember he'd wanted to build things, and a move to Charlottesville would give him a chance to earn an engineering degree at the University of Virginia. Since he hadn't majored in engineering as an undergrad, he'd need to take several crash courses in advanced mathematics.

When Oliver arrived at Estouteville, he found the mansion in a state of chaos. Beatrix had launched a major historical renovation, and soon she would begin transforming parts of the estate into an elaborate gallery for her own paintings and installations. She built labyrinths out of bushes. She painted a felled tree blood red and hung bird nests from the branches. *Nature Morte*, she called it.

Amid all this hubbub, Oliver found it difficult to do his homework. Because of the renovation, he and his parents had to share a small guesthouse in back. It was crowded in there. Oliver couldn't study. He fell behind in his classes.

After a few months of this, it hit him: If I need to engineer something, I'll just hire an engineer.

6

Engineers

Up-angled 600 feet into a chlorinated blue sky, he heard the noise—a burble in the exhaust of the little plane's two-stroke engine. Maybe there was too much fuel in the combustion chamber, he thought. He put his hand on the throttle and moved it a quarter inch to try to fix the problem. Then there was another noise. Then there was relative silence.

Oh shit, thought Barnaby Wainfan.

The engine had cut out, stranding him inside his own invention. The plane coasted for a moment or two, bleeding velocity. Then Barnaby pushed the nose down, to maintain airspeed and begin a glide.

He was above Blythe, California, a desert region halfway between Los Angeles and Phoenix. Barnaby, a 39-year-old aerospace engineer with a rabbity build, darting eyes, and glasses, could see a gas station ahead of him, a highway off to the right, and lots of arid land spotted with sagebrush. He had to make a decision: Try to land or deploy the chute? The plane, which was about as light as a single Harley-Davidson motorcycle, was equipped with a whole-vehicle parachute. Barnaby thought about using the chute, but when you do that, you're landing wherever the wind takes you. You're also guaranteed to destroy the plane.

He was loath to do that. He'd put three years of his life and four thousand hours of work into the plane's design. He called it the Facetmobile, or sometimes "my hallucination." It looked like a stealth fighter, one of those matte-black wedges, but radically simplified. The entire plane was wing—a folded sheet of eleven flat geometrical planes, or facets. Built from aluminum tubing and a lightweight shell of white polyester fabric, it could be assembled by a home hobbyist on a couple of workbenches. The promise of the Facetmobile was the democratization of flight: Finally, every American of modest means could have a little stealth fighter under the tarp in his backyard.

Barnaby opened the throttle, and the engine came back online at a low level—not enough to fly the plane properly, but maybe enough to land it safely. He decided to turn back toward the small airport from which he had just taken off and attempt a landing.

He managed to get the plane on a runway heading. Time slowed down; the rest of the world went away. Through the windshield, he saw the dirt before the runway, a barbed-wire fence, the runway on the other side. And the propeller turning and turning.

Ten seconds later, the engine went cold.

He was traveling 70 miles per hour. There was no control tower at the airport, hence no air-traffic controller, but Barnaby got on the mic anyway. "Mayday Mayday!" he cried. "Experimental 7 Whiskey Delta, engine failure, landing 3-0 Blythe." No one was listening. All he could hear was a little bit of wind over the skin of the plane and a loud stream of profanity. He cursed the engine all the way down.

Just above the ground, Barnaby suddenly realized he was going to come up short of the runway. He aimed for the dirt. The final few moments before impact were deceptively peaceful. He managed to get the landing gear on the dirt without dragging the butt of the plane. Then a loud *ka-flap* smacked through the cabin. He had crashed into the barbed-wire fence. The plane rolled to a stop. Barnaby realized that he was on the ground and unharmed. He thought, *I got away with it.* Then he saw a blur of rubber and dust off to his left. It took him a second to understand that the blur was the nose wheel, part of the plane's landing gear, bouncing away in the dirt, like a disintegrating jalopy in a Bugs Bunny cartoon.

TODAY, SEVENTEEN years later, Barnaby insists that what he experienced in Blythe, California, in 1995 was not a crash but "an engine failure–induced mishap." When he landed, he was in control of the plane. "If it hadn't been for the damned three-wire barbed-wire fence in the way, it wouldn't have been much of an incident," he says. Still, "as I walked away, it hit me—holy shit, I coulda died, you know?"

Mostly Barnaby was just pissed off that he'd broken his plane, the fulfillment of a dream that had gripped him since he was five. He had grown up in Queens, New York, in what he describes as "an intellectually intense household"; his mother was a biochemist, his father a physics professor who sat on Barnaby's bed at night and read to him from Jules Verne. Barnaby's favorite book was Verne's *The Mysterious Island*, the tale of five men who crash a hot-air balloon on a crag of rock in the South Pacific. The men go on to reinvent civilization using nothing but the things in their pockets and the natural materials of the island, starting with an object called a "burning-glass." One of the men, a railroad engineer named Cyrus Harding, takes the glass faces of two pocket watches, seals them with some clay, and focuses the rays of the sun through the glass onto dry moss. The moss bursts into flame. Another castaway "considered the apparatus," Verne writes. "Then he gazed at the engineer without saying a word, only a look plainly expressed his opinion that if Cyrus Harding was not a magician, he was certainly no ordinary man."

As Barnaby sat up in bed, spellbound by the ingenuity of Verne's heroes, he was seized by a conviction that would never quite leave him: *This is how people should be.* Building stuff was an ethical act, a way to become more fully human.

"The Facetmobile," he crowed four decades later, "is for people who fly because it's good for their soul."

Even after the crash, Barnaby's street cred in the DIY aviation community kept soaring; *Popular Science* called him one of America's "daring visionaries of crackpot aviation." Crackpot was a "most unfortunate" word choice, as Barnaby pointed out in a subsequent letter to the editor. "Inventors like those profiled in your article," he wrote, "follow personal visions that drive them to persist where large organizations give up. It takes a deep, almost illogical level of optimism to keep pushing forward in the face of skepticism and the technical and financial obstacles inherent in advancing the state of the art. Many fail, but those who succeed enrich us all."

The article was inaccurate in another sense. Describing Barnaby as a visionary of aviation hardly captured the man's essence, the way he busted out in an escape pod whenever he felt himself too confined in a particular ship. This was a guy, after all, who had built everything from a model submarine made of wood and tin cans that traveled the length of a pool underwater (he did that when he was twelve) to plywood boats that he paddled around the local harbors with his three daughters. He also made model planes that he flew competitively against other planes in a sport called control line combat—graceful, slow-moving aerial dogfights to the death. Barnaby's model planes were so cleverly constructed that, more than once, the control line combat judges were forced to revise the rulebooks to explicitly ban Barnaby's techniques. He interpreted this as "the highest possible compliment" to his skills as an engineer.

Barnaby's great strength was his fearlessness; he wasn't shy about building things that might look ridiculous or might never work. In 2001, on an episode of the reality show *Junkyard Wars*, Barnaby helped a family from Oregon hack together a glider from scratch in ten hours. Then he put on a lime-green helmet, climbed into the glider, pointed it down a hill, and achieved about three seconds of flight before pitching and rolling into the dirt. Barnaby claimed afterward, not without pride, "I designed it *assuming* it would crash." And in 2008, on assignment for a small military contractor called Atair Aerospace, Barnaby built a loopy contraption called the EXO-wing, a twin microturbine-powered jet plane so tiny and lightweight you could wear it on your back. It was the world's smallest human-piloted jet.

And all of this was strictly extracurricular. In his day job, he designed multibillion-dollar weaponry. Barnaby was a technical fellow at Northrop Grumman, the megalithic military contractor. His security clearance was top secret. One project had him working as the lead aerodynamic designer and lead flight-mechanics engineer on the team that gave birth to a completely new kind of vehicle, Northrop's X-47B, a drone that could catapult-launch itself from the deck of an aircraft carrier, drop a bomb 2,000 miles away, and return to the carrier, piloted entirely by artificial intelligence. (Currently, the drones used by the U.S. military in Pakistan and Yemen have to be controlled by humans, and they also have to take off from the ground, which is why a carrier-based drone is desirable; the military wouldn't have to build airstrips in foreign countries.) Many of Barnaby's other Northrup projects are classified.

Asked if he has ever designed things that have killed people, Barnaby doesn't hesitate. "Yes, absolutely," he says. "I designed things that saw combat in Operation Desert Storm."

Is he okay with that, morally?

"Have you ever seen the Patrick Swayze movie *Road House*?" he says. "Well, the basic philosophy is, you need to be nice, until it's time to *not* be nice. And when it's time to not be nice, you *end it*."

How to boil down Barnaby Wainfan? A man who quotes eighties B movies, has studied karate, and has written blues songs about thermodynamics? Who seems to know everything about everything and wants to tell it all to you in one panicked breath? Some engineers who've known him forever have come to dread being stuck with him in meetings, knowing that they'll never get a word in, but to someone who has never met him, Barnaby is the most fascinating person in the world. (His old acquaintances used this to their advantage. "We used to have this thing we called 'throwing a virgin to Barnaby,'" says a former Boeing engineer who moves in the same private-aviation circles. "We'd find someone he'd never met and say, 'Hey, come meet Barnaby Wainfan!' And then we could go off and talk about something else.") He's clearly a quirky character, but beyond that, at the root of the Barnaby conundrum, is the sheer range of his output—where does one guy get off asserting his mastery of planes, boats, bombs? Barnaby has made stuff that's small and frugal and joyous, but he has also made stuff that's staggeringly complex and kills people—and yet when you ask him about it, he shrugs and says it's all the same thing, really. "Air is air," he says.

Air. An invisible common thread. While other people perceive planes and bombs as solid objects moving through empty space, Barnaby has trained himself to perceive air as a complexly sloshing fluid perturbed by the objects moving through it. Most people wake up in the morning and see a world of stuff. Barnaby sees the inverse, a world of jiggling wakes—in air, in water. Barnaby takes for granted what we often forget: that air isn't empty. In fact, from the point of view of an object at speed, the difference between water and air is mainly one of density, of chili versus chicken broth. Either way, if you're an object that wants to move through water or air, you first have to move aside the soup that's directly in front of you, which takes a certain amount of energy. It's always nice to use less energy to do the same amount of work, and this is where an aerodynamicist like Barnaby comes in.

There aren't many aero designers in the world, simply because there aren't many new planes or boats or cars released in a given year. Northrup Grumman has 120,000 employees and twenty or thirty aerodynamicists, three or four of whom are masters. "It's always been the case that you have a few masters of the art," Barnaby says. "I would hazard a guess that, within the U.S., there are probably five or six people who can do it at the level I do it. Maybe ten. And if there are more than a couple of dozen in the world, I'd be surprised."

In 2007, the engineer Ron Mathis happened to bump into Barnaby while waiting for a flight at LAX. They were strangers, but they talked for an hour straight. Mathis later told his friends that Barnaby was one of the most interesting people he'd ever met.

RON WAS the opposite of Barnaby in some ways: a ropy vegetarian with a voice so deep and so soft you had to strain to hear it. He was a gun for hire. He helped rich men and major automotive companies win car races. He had his own plane—a lightweight, twin-engine Beechcraft Baron—and flew it around the country, from job to job. He enjoyed the challenge of operating alongside the big planes; while overripe 747s darkened the sky, Ron Mathis, small and light, squeezed between them.

Over the years, he'd worked on teams that raced heavily modified Porsches, Lotuses, Ferraris, Fords, and Audis on road courses full of hairpin turns, often in endurance races that lasted up to twenty-four hours, stressing every component to the point of failure. His duties shifted depending on the car and the team. Sometimes a team owner—the automotive company or the wealthy

individual signing the checks—would ask Ron to design and
build a car from scratch, as "chief of design." More often, though,
he came into the picture after the car was already designed and
built. In this case, Ron was a "race engineer." He would direct the
mechanics on how best to make the car glide through straight-
aways and corner sharply. He carried at all times a metal clipboard
containing several clickable pens and a scratch pad, along with his
laptop, which was loaded with CAD software and some mathematical models he had written. He studied curves of drag, speed,
and thrust. He worked with the car before the race like a horse
trainer works with a thoroughbred before the Kentucky Derby.

He'd been refining his eye for automotive detail ever since he
was a kid in southern England, growing up the brother of a man
who raced cars for fun and the son of parents who sold cars for a
living. ("If you look at my family photo albums," Ron says, "you see
mostly pictures of cars. People just happen to be in them from time
to time.") He was the kind of person who liked to look at sports cars
driving in the rain, because he could tell whether the aero of a car
was good or bad by looking at the trails left by the raindrops and
the spatter patterns of mud. Careful, dry, and direct, he had earned
a reputation in the industry as a sort of monk who had consecrated
his life to his art. Individuals and companies blowing millions of
dollars on something as extravagant and risky as high-end sports-
car racing liked the purity of his approach. Racing could be silly,
deadly, and wasteful, but the way Ron went about it, it was also a
heroic quest for perfection. "In the industry, he is like a god," Oliver tells me, "but you've never heard his name."

Ron made a decent living for a time, bouncing between teams
based in Tennessee and Ohio and consulting on the side. He
bought a house two miles from the Indianapolis Motor Speedway
and lived there with his wife and teenage son. Starting in 2005,
Ron got a job with Audi's racing division, training a new set of
horses: Audi R8s and R10s. These were some of the finest race cars

ever constructed, true thoroughbreds, and people who knew what they were drove hundreds of miles just to see them run.

The Audis racked up win after win in prestigious endurance races. Despite his success, though, Ron was growing weary of the racing world. In order to make the races close and entertaining, and therefore attractive to advertisers, many of the circuits required cars to conform to increasingly rigid specs, robbing engineers of the freedom to make substantive changes. "The nature of the process is that you have to become obsessed with details," he explains. "Little things here, little tweaks there, some little widget, because that's what you're allowed to do." It used to be that the goals of racing and consumer R&D were one and the same—to make better road cars. Now the automakers' consumer divisions were searching for the holy grail of fuel efficiency, while the brilliant engineers in their racing divisions were making tweaks to the latest gas-guzzling V8s. It was an absurd waste of human capital, as if Silicon Valley's most elite programmers had only spent the last two decades optimizing video-game code instead of creating search engines.

In 2008, when the global economy crashed, the rich recoiled from racing, and many car companies, looking for a quick way to cut costs, disbanded their teams. Ron was laid off from Audi along with all of his colleagues. "One of the best racing teams I'd ever seen just scattered to the four winds," he says.

A few weeks later, Oliver called.

Ron had worked with Oliver before on car projects, but Oliver had never been the guy calling the shots, only the rich guy writing the checks. Now Oliver was talking about a different arrangement. He said he was assembling a team that he was going to lead himself. He'd heard about a $10 million prize for fuel efficiency and he thought racers could win it.

So much of racing was about mastering drag and shaving weight. Most of the time, that knowhow was used to eke out a couple more revolutions of the engine from a drop of gas, in a

car burning gas at a rate of 1 gallon every 3 miles. But what if racers used what they knew to make a car that sipped gas instead of
gulped it? Could Ron come to Charlottesville and talk it over?

When Ron arrived, the first thing he and Oliver did was go for
a ride in Oliver's new Volkswagen Jetta TDI sedan. They wanted to
establish a baseline: What's the upper end of current engineering?
Can we modify an existing car to get 100 miles per gallon, or will
we have to go in a different direction? As Oliver steered through
local hills, the Jetta flashed its MPG in big fat digital numerals. The
figure changed by the second. Going uphill, Ron felt the engine
kick in, working extra hard to drag its 3,300-pound bulk and four
hundred pounds of human flesh to the crest. The MPG number fell
into the twenties. Flying downhill, the number rose to 90, 100, 103,
104, but it always dipped back down again. No matter what Oliver
and Ron did, they couldn't crack 100 for more than 1 or 2 seconds.

Before coming to Charlottesville, Ron had scanned the X Prize
rules. He saw that there was no easy way to win. Like "real" cars,
the Prize cars would have to demonstrate competency at a wide
variety of tasks: acceleration, braking, cruising, starting and stopping, hill-climbing, durability, range. The trouble was, designing
a car involves trade-offs. If you add a bigger engine to increase
acceleration and hill-climbing ability, you decrease your efficiency
and your range. It was a classic engineering puzzle. How do you
make a car that is powerful enough to accelerate and climb but
not so powerful, or so large, that it can't get to 100 MPG? To Ron,
the solution now seemed clear: a combination of a small, efficient
engine and a frame and body that were lighter than those on any
production car ever made.

CONCEPTS FOR an ultra-light car had been floating around for
decades. In the early 1990s, a Colorado environmentalist named
Amory Lovins proposed a class of lightweight, hybrid-drive vehi-

cles that he called Hypercars, claiming they would reduce the need for oil so drastically that "the Middle East would therefore become irrelevant." The Hypercars would be light because they would be made of advanced composite materials reinforced with carbon fiber. But despite twenty years of promises and $6 million in capital, Lovins never managed to build a single drivable prototype.

General Motors got further. In 1991, three GM vice presidents asked one of their engineers, Jim Lutz, if he could build a super-efficient concept car for the 1992 Detroit Auto Show. Lutz and his team set up their own lab and machine shop at GM's Technical Center in Warren, Michigan. Over an intense four-month period, working nights and weekends, they put together the Ultralite, a functional prototype built around a composite chassis fabricated by Burt Rutan, the aerospace pioneer and future winner of the Ansari X Prize. Featuring a low-drag shape with a small rear-mounted engine and gullwing doors, the Ultralite weighed 1,400 pounds and achieved 100 MPG when running at a steady 50 miles per hour, a much lower bar than the X Prize cars would later have to clear. (The Prize cars couldn't just excel on highway driving; they would also have to make quick stops and starts, mimicking city driving, and their performance on the city portion would count toward their final MPGe.)

After its debut at the Auto Show, the Ultralite made the cover of *Car Design*, *Automotive Industries*, *Materials Engineering* (IS IT FOR REAL?), *Popular Science* (DETROIT DESIGNS FOR THE FUTURE), and *Popular Mechanics*, which wrote that the Ultralite "is for real" and that "General Motors has never been more serious." Lutz gave rides to GM board members at a desert proving ground in Mesa, Arizona, and during the filming of *Demolition Man*, the 1993 Sylvester Stallone action movie set in the future, he put hard miles on the car in chase scenes. As far as Lutz knew, GM wanted to keep moving forward and evolve the car into a production model, so he had Burt Rutan build a second Ultralite chassis for use as an engineering "mule" vehicle—the next step in the development process. But that's as far as things got. Soon the GM vice presidents who had championed the project were reassigned in a management shakeup, and the new bosses saw the Ultralite as a stunt, not a real car. They junked the program to cut costs. Lutz's lab shut down and his team disbanded.

More recently, Toyota and Volkswagen had introduced lightweight prototypes at auto shows. Toyota's, the 1/X, weighed in at 926 pounds, and Volkswagen's, the 1-Liter, was a mere 639 pounds. But, like the Hypercar, the cars were light because they were made of materials like carbon fiber, magnesium, and titanium, which are incredibly stiff and strong and light but also more difficult to work with than steel and aluminum and therefore more expensive. Oliver not only wanted his car to be inexpensive, he wanted it to dissolve easily into the global scrap-metal stream. Each year, about 14 million cars are junked in America. His car, he envisioned, would be almost completely recyclable, the death of one car giving birth to part of another in an endless cycle, a concept known as "cradle-to-grave sustainability." (Oliver preferred to call it "dust-to-dust manufacturing" because he liked the way that sounded.) The car had to be made largely of cheap, widely available materials like aluminum and steel. Most cars are largely made of steel, but Oliver and Ron would use a greater proportion of aluminum, which is lighter than steel and, used well, can be just as strong.

Having committed to making a light car, Oliver and Ron now needed to pick a fuel. Oliver drove back into downtown Charlottesville and took Ron to the Skatetown Building, so named because it had been a roller skating rink before Oliver bought and renovated it. Ron walked behind Oliver along the side of the building, on an overgrown, weedy path that led to an unmarked door. Through the door was a room full of flyers for a bus service called the Starlight Express.

Starlight, which Oliver owned with a friend, had been his

first attempt at putting his stamp on the transportation industry. As Oliver tells it, he'd always felt that a bus would be a pretty reasonable way to travel if only you made the ride more comfortable. So he'd bought an old Trailways bus on eBay, ripped out all forty-eight seats, and installed twenty-four beige leather seats he had spent a year scavenging from BMWs. Then he plastered the inside of the bus with posters of John Lennon and Yoko Ono. The Starlight drove a single route: Charlottesville to New York City. Sometimes Oliver drove the route himself—a multimillionaire clocking in as a bus driver. The flyers, which Oliver had written himself, said:

WHO RIDES THE STARLIGHT?

Filmmakers ✳ Engineers ✳ Vietnamese

Swamis ✳ Dentists ✳ Bosnians

Mothers ✳ Monks ✳ Israelis

Palestinians ✳ Tattoo Artists

Ukrainians ✳ Runners

Ron noticed that the Starlight office had a postal scale. He was starting to think about the weight of various fuels. There was no point in designing a very light car that ran on a heavy fuel. He needed a fuel that was as energy-dense as possible, that had the most bang for the ounce. He and Oliver ticked through a list of candidates. Compressed natural gas? Maybe, but it required a big, bulky tank that would mess with the aerodynamics of the vehicle. Hydrogen? Same problem. The least dense of all fuels. Hydrogen had other downsides, too: It was highly explosive, it took a lot of energy to make, and there was no refueling infrastructure.

What about electric batteries? Mathis turned his laptop upside down. It was a newish Dell, and Ron figured that its lithium-ion battery—the same kind used in many electric cars—was close to state-of-the-art. He removed the battery and placed it on Oliver's postal scale.

Lithium is the lightest metal in the universe, about half as dense as water. This is why batteries made of lithium are lighter than traditional lead-acid batteries. Still, they're not as light as you might expect; most of the battery is made of inactive ingredients like plastic packaging, copper (which carries the charge), and electrolytes (the solvent that separates the electrodes). Ron's laptop battery had a number written on it that told him how much energy it held. Once he knew the weight of the battery, he did a quick calculation. He found that he would need at least 500 pounds of lithium batteries to power a very light car, at which point the car would no longer be especially light.

So Ron came back around to good old gasoline, partly because it was what he knew, and partly because there was no way around it: Gasoline is remarkably energy-dense. There's more juice in a cup of gasoline than in a cup of any other common fuel except nuclear fuels like uranium and plutonium. In 1896, when Thomas Edison met a young Henry Ford for the first time, at a dinner near Coney Island, Ford grabbed a menu and sketched his idea for a car that ran on gasoline rather than electricity or steam. (Ford hadn't invented the internal-combustion-engine car—a German, Karl Benz, had done that—but at the time, there were only a handful of gasoline-powered cars in America.) Others had struggled to make a viable electric car, but the batteries were too bulky, too unreliable. Edison told Ford, "Young man, that's the thing! Electric cars must keep near to power stations. The storage battery is too heavy. Steam cars won't do either, for they have a boiler and fire. Your car is self-contained—carries its own power plant—no fire, no boiler, no smoke, and no steam. You have the thing. Keep at it." Ford convinced the American public to believe in gasoline cars. And then Ford became Ford, and America became America.

Having settled on gasoline or some other kind of liquid fuel—maybe a gasoline-ethanol blend—Oliver and Ron drove Oliver's pickup truck to a motorcycle shop called Jarman's Sportcycles. They walked up and down the rows of candy-colored bikes and came to a stop at a small, shiny, gasoline-burning Yamaha. The Yamaha was the sort of bike that motorcycle magazines recommended for beginning riders. A starter bike. When it idled, it went *putt-putt-putt*. The volume of its single engine cylinder was a miniscule 250 cubic centimeters, or about two-thirds of a can of soda. But Oliver and Ron reasoned that in a very light car, a soda can–sized engine was all you would need, as long as the aero was good.

At highway speeds, a car spends more than half of its energy just pushing the air out of the way. The less pushing it has to do, the more efficient it can be. Ron calculated that for the Very Light Car to hit 100 miles per gallon in the Prize, it would need a drag coefficient of less than 0.20. A drag coefficient is a quick-and-dirty measure of a car's sleekness. The lowest-drag production car in history was the EV-1, the ill-fated electric car from General Motors, with a drag coefficient of 0.195. No production car had ever beaten that. For the Very Light Car to win the Prize, it had to be sleeker than the sleekest production car in history.

At Ron's suggestion, Oliver called Barnaby Wainfan and told him about the Prize and the short time frame. Edison2 had to reinvent the automobile in less than two years. This left no time for extensive testing—no wind tunnel, no computer simulations. The aero of the car would have to be eyeballed. "I need someone who can get it right enough to win, by eye, on the first try," Oliver told Barnaby. "And Ron says you're that guy."

"No guarantees," Barnaby replied, "but I'd love to have a shot at it." Barnaby didn't know much about cars, but he knew about everything else. Airplane fuselages. Bomb shapes. Blimps.

Zeppelins. "There's a huge body of information out there that hasn't been used for cars. But the physics of the air is the physics of the air. It hasn't changed." The more Barnaby thought about it, the less daunting it seemed. He called Ron. "Ron," he said, "if the first number after the decimal point isn't a one, you'll have to take my Aero Man card away."

In California, Barnaby changed his daily routine. He drove to Northrup, as always, in his silver Grand-Am with a row of tiny squares of scrap aluminum taped above the rear window: home-made "vortex generators" he had added to quiet the air as it flowed along the car's afterbody. He spent the day working for the second-largest military contractor on the planet. Then, at five, he drove to a coffee shop in a strip mall, near his karate dojo, to sketch. "I always start by sketching," he says. "There's an aesthetic element to the whole thing. One of the things I love about doing aero design is that it's still *art*."

When most people picture someone designing a plane or a car, they see computers, sheets of polished chrome, glass-windowed labs. But Barnaby sees it more like jazz. To be a good jazz player, you need to master the basic techniques of music theory and tone—how to play scales, how to strum a guitar. "But when it's time to perform," he says, "and you're improvising, and stepping up to the mic, and letting it fly, you're using all that stuff, but that fundamental sound is not derived by analysis."

Barnaby ordered a cup of Chinese black tea. He sat at his favorite table, between posters of Stevie Ray Vaughan and Ray Charles. He flipped open a pad of graph paper, a no. 2 pencil in his hand, and began to sketch.

HE STARTED with a couple of assumptions. One was that Edison2 couldn't afford the "open bottom" you find on a typical car, with all that raggedy stuff—muffler, oil pan—hanging out and snagging

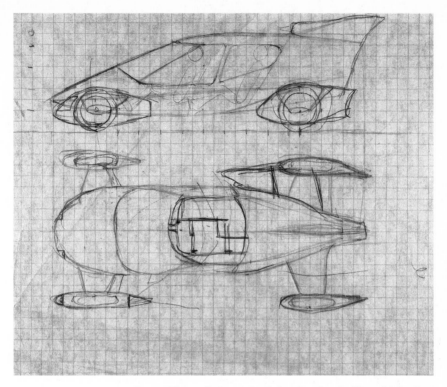

One of Barnaby's original sketches for the Very Light Car.

the air. The Very Light Car had to be well sealed, its underside as smooth as a dolphin's belly. Another thing he knew was that the wheels couldn't be situated within the car like traditional wheels— flush with the body and exposed to the air. To the air, a tire looks like a cylinder, and a cylinder is 5 to 10 times less aerodynamically efficient than a streamlined shape like an airplane wing. Also, when air zips through the open fender wells of a typical car, it gets stuck in the holes and burbles, generating turbulence, which is lost energy, lost MPGs.

Barnaby thought about airplanes with landing gear that doesn't retract; the wheels are often covered with "fairings," curved shapes like miniature wings. He figured he could enclose the Very Light Car's wheels and axles within similarly shaped fairings and extend them away from the body. The wheels, in effect, would have their own wings. The car's stance would also become wider, improving stability and making it handle better. Handling was important, because, at the time, the X Prize was supposed to be a multi-city race; the car would have to navigate a lot of turns, and if the car handled well, the driver could keep the speed up around the turns, conserving energy.

One disadvantage of putting the wheels "outboard" of the body is that parts of the suspension would stick out, catching the air. You could always wrap them in fairings, as one of Edison2's chief X Prize competitors, Aptera Motors of California, had done with its vehicle, the 2e, whose front wheels were outboard. But Barnaby found that solution inelegant. He and Ron talked about it, and they decided to find a way to confine the suspension to the wheels. It would free up all kinds of space in the front and the back of the car, but especially the front. As a result, the car wouldn't need to have as much "frontal area"—the part that smacks the wind head-on. The front could be lower, more tapered.

Barnaby pondered a quality called "base drag." In a perfect world, the car would part the air, the air would come back together around the car, and the air would never know the car was there. The air would perfectly fill the hole that the car had just punched in it. Typical road cars are bad at this; they split the air in the front and neglect to zipper it back together in the rear, like an airplane does. They leave rough holes in the air. They generate base drag. Barnaby considered two kinds of zippering shapes: a "beaver-tail," which splits the air vertically (some air goes underneath the car, some goes over the top), and a "boat-tail," which parts the air horizontally—the air goes to the left and the right, like pedestrians on a crowded city street navigating around a square of wet concrete. Most cars aren't aerodynamic enough to qualify as either a beaver-tail or a boat-tail; as far as the air is concerned, they might as well be couches.

The Aptera 2e is a beaver-tail design. But Barnaby preferred the boat-tail approach, because a beaver-tail creates a channel of air between the car and the ground that can get turbulent; the air has to squeeze through a small space beneath the car and then rapidly climb up the back of the car to meet the air flowing over the top. Boat-tails are difficult to implement using traditional car designs, because the body has to enclose the rear wheels; this means the car has to be ridiculously long to keep the slope of its body gentle enough for good aero. But the Very Light Car wouldn't have this problem, because its body would be separate from its wheels.

In the coffee shop, Barnaby sketched several boat-tail architectures. One resembled a catamaran: a four-seat car divided into two pods, connected in the middle by a slender piece of aerospace composite, almost like two covered motorcycles running in formation.

After thinking about it for a day or two, Barnaby rejected the catamaran: "Most people will look at this thing and say, *That's just too weird.*" Also, in a crash, the two pods might fly apart in different directions.

"I will lay down three or four concepts that range from pretty conventional to pretty radical," Barnaby says. "I will sketch an idea over and over and over again. Sometimes it looks like: *Man, he's drawing the same thing ten times.* But if you look carefully, each one is very subtly different."

Searching for inspiration, Barnaby scanned his mental database for boat-tailed cars. When he was a kid, he'd devoured a book called *The Golden Age of the American Racing Car*, particularly the section about a working-class mechanic named Frank Lockhart. The book described him as a "little guy who slew giants by virtue of nothing but guts, infinite labor, and, above all, brains." In the 1920s, sponsored by the Stutz Motor Company, Lockhart built an unusual car called the Stutz Black Hawk Special. It had a completely flat belly. The wheels stuck out from the body of the

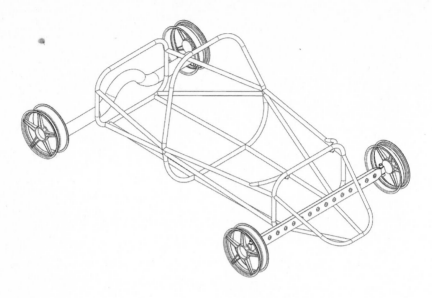

An early CAD sketch by Ron.

car and were covered with aerodynamic fairings. Using an engine
built by Harry Miller, a genius race-car designer from Wisconsin
whose machines won the Indianapolis 500 nine times, the Black
Hawk was so efficient that Lockhart was able to pursue a land-
speed record using an engine 6 and a half percent the size of the
previous record-holder. One afternoon in 1928, while he was mis-
siling across a stretch of Florida beach at 220 miles per hour, one
of his tires popped. The Black Hawk flipped wildly end over end,
flinging its driver 51 feet across the sand, killing him instantly. But
it was just bad luck. Anyone can blow a tire.

Ron knew about the Black Hawk, too, and he and Barnaby
kept coming back to it. "We both kind of said, well, the whole
approach makes a lot of sense," Barnaby recalls. They also discussed an English sports car called the Ariel Atom, "a hideously
powerful car with an open-tube frame—a scary beast." Barnaby
told Ron that what they needed was "basically Buck Rogers meets
the Ariel Atom": a chassis like the Atom's with a retro–science

fiction shape wrapped around it. Ron understood immediately: "Yes," he said, "that's it, let's do that."

Barnaby e-mailed his sketches to Ron in Lynchburg, who opened the CAD program on his laptop and began to draw the lines and curves of a frame—his best guess based on Barnaby's shape. It featured tubular side trusses joined by three bulkhead hoops. Ron sent the CAD file to Barnaby, who used it as a guide when he made his own 3-D model. Ron's frame stuck through Barnaby's surface in a few places, so Ron revised the frame several times to fit Barnaby's work, and then Barnaby revised the surface to fit Ron's frame. After two or three iterations, the engineers had agreed on the basic profile of the Very Light Car. It harkened back to the Black Hawk and to Miller's Indy cars. The wheels and the axles that connected the wheels to the car were "faired," like the Black Hawk. The back end of the car terminated in a hard vertical line that resembled the bow of a boat. The engine would be mounted in the back instead of the front.

Barnaby also had to find a way to control the car's cooling flow; its little soda can–sized engine would be working hard, like a hummingbird, generating heat that the car would have to dissipate, and a rear-engine car is inherently harder to cool than a front-engine car because you have to find a way to route air from the front to the back. So he designed an inlet on each side of the car, two-thirds of the way toward the rear, to direct air toward the engine and the radiator. The inlets were scooped into the car and triangular in shape, with the point of the triangle toward the front. This type of inlet is known as a NACA duct, named after the National Advisory Committee for Aeronautics, the forerunner of NASA. Originally invented as a way to cool planes without increasing drag, NACA ducts are also used on race cars. Barnaby tried to keep the inlets small, because the larger the inlet, the more it would affect the car's drag. It was a tricky thing to pull off, because the inlets were *behind* the widest portion of the car, which meant that to reach them, the air would have to make a slight turn. It would have to stay "attached" to the car, to scoop around the horizontal bulge in the middle and follow the contour of the car, and then flow into the inlets. If Barnaby didn't get the shape of the car exactly right from an aerodynamic standpoint, the air would detach from the car, come unstuck, and never make it into the inlets. And then the car might overheat.

Barnaby and Ron didn't worry that the shape of the car might strike some people as strange. Nobody who boards a plane questions why a plane looks the way it does, they reasoned. The plane looks the way it does because that's the price of being light enough to get into the air.

7

Yoda and Little Old Man

n late 2008, while Edison2's Barnaby Wainfan noodled jazz solos in California, another X Prize competitor was obsessing about the air: Kevin Smith, the government bureaucrat in Illinois. Kevin didn't know the first thing about aerodynamics. Nor did he have the money to hire his own Barnaby Wainfan.

What Kevin had was an old aerodynamics textbook that he'd read about online. Called *The Leading Edge*, it was so obscure that no copies were listed in any of the major online bookstores. Kevin managed to get one through an interlibrary loan, and the day it arrived, he immediately removed the book from its manila envelope, skimmed it, and made photocopies of the relevant pages so he could look at them during work hours at the Illinois Environmental Protection Agency and make it seem like he was doing actual work. He stashed the copies in a folder at his cubicle, beneath a shelf full of *Star Wars* memorabilia and a cloth wall lined with *Dilbert* cartoons and inspirational quotations, including the famous one from Thomas Jefferson about how a government big enough to supply everything you need is big enough to take everything you have. There was also an anonymous quotation that reflected Kevin's somewhat sick sense of humor and drew strange looks: "Make a man a fire and he's warm for the night. But light him on fire and he'll be warm for the rest of his life."

Kevin was able to knock out his permits in a couple hours, leaving the rest of the day to pore over the aero book. At 6 P.M. sharp, he left work, climbed into his Ford pickup truck, and drove the twenty-seven miles to his home in the farm country surrounding Divernon, carrying the photocopies of *The Leading Edge* with him. The sun set in the cornfields—the same cornfields Charles Lindbergh had flown above in his biplane eighty years before, while carrying the mail from St. Louis to Chicago. Kevin's wife, Jen, came home from her graphic-design job and made Kevin dinner. Kevin stared at the long, involved equations and weird terminology in the aero book—*bluff body, viscous friction, boundary layer, superlayer*—and tried to make sense of it. If he couldn't master the aerodynamics equations, he couldn't build an aerodynamic car. And if he couldn't build an aerodynamic car, he'd never win the X Prize. And if he couldn't win the X Prize . . .

Well, but he could, right?

Isn't that what Paul Cox, the old man back in his hometown of Park Forest, would tell him if he were still alive? Didn't he owe it to Paul, to the memory of Paul, to try?

PAUL G. COX, who stooped when he walked and taught Smith to chase possibility. Paul, who devoured cheap-ass books and played chess with bawdy old men at the South Chicago Social Club. Paul's history had always been obscure to Kevin—there were parts of his life Paul just didn't talk about. But over the years, Kevin did manage to glean a few key facts about the man who had become his most important mentor.

As a young man, Paul had studied philosophy. Then he was drafted into the army and served on a U.S. military base in Oki-

nawa during the latter part of World War II. It was unclear whether
Paul had fought in the Battle of Okinawa, one of the bloodiest of
the war, but in the army he wasn't a man of rank, just a grunt, so
it was possible he had seen combat. (In later years, the only story
Paul told Kevin about Okinawa had to do with a knee injury he
sustained playing handball on the base.) After the Japanese sur-
rendered, he took a picture of the surrender plane flying over Okinawa, headed back to Japan.

When he returned to the U.S., what Paul wanted, like so
many other GIs, was peace and quiet, a nice place to raise a family.
Around 1955, in one of the Chicago newspapers, he saw an adver-
tisement for a place that seemed perfect. It might well have been
an ad like this one, which appeared in 1952:

> YOU BELONG IN PARK FOREST!
>
> The moment you come to our town you know:
> - You're welcome.
> - You're part of a big group.
> - You can live in a friendly small town
>   instead of a lonely big city.
> - You can have friends who want you —
>   and you can enjoy being with them.
>
> Come out. Find out about the spirit of Park Forest.

The village of Park Forest, thirty miles south of Chicago, was
created from scratch by a visionary named Philip Klutznick. During World War II, Klutznick had worked for the U.S. government
as a housing guru, supervising the construction of several instant
cities for all the defense workers hastily assembling tanks, ships,
and ammo. Oak Ridge, Tennessee, where the first atomic bomb
was built, was a Klutznick creation. He later said, "I don't know if

the atomic bomb could have been built if it weren't for the expand-able trailer." After the war, Klutznick sought to apply what he had learned to peacetime communities.

Today, the centrally planned, automobile-lubricated suburb is everywhere, but back then it was a new and wacky idea, and Park Forest, which opened in 1948, was one of its first incarna-tions. (Construction on Levittown, New York, considered to be the first postwar suburb, had only begun one year before, in 1947.) Klutznick believed that people would trade proximity to their job for comfort at home. They would own cars and drive them to work and back home again. Klutznick built the roads of Park Forest, built the townhomes, built the shopping center. He lured young families from Chicago by promising the good life.

Paul Cox came to Park Forest with his wife, Monica, when it was still a town of mud. They moved into a house with powder-blue cedar siding on a quiet, not-yet-tree-lined street. They had three sons and sent them to Park Forest's schools. Paul grew old in Park Forest, and his body deteriorated. After twenty years in the house, he could no longer walk normally, thanks to his old hand-ball injury. If he stood in one place too long, he'd have to hit the back of his knee to get it to bend so he could start walking again.

In 1978, a grocery-store manager named Nick Smith rented a house around the corner from Paul's. He had three children, two boys and one girl. The middle child, Kevin, was thirteen when he started mowing Paul's lawn to earn some extra money.

One Sunday, after Kevin had finished with the lawn, he went inside to collect his cash, and for some reason that Kevin will never understand but will always be grateful for, Paul said, quietly, "Do you play chess?" When Kevin said he did, Paul invited him to play.

"I'm rather good," Paul warned him.

"So am I," Kevin said.

Paul beckoned Kevin into his living room. It was splotched

with old throw rugs and rickety furniture that Paul hadn't replaced for thirty years. Paul slumped toward a folding table. He lowered himself gingerly into a chair. He plucked a set of plastic rooks and pawns from a satchel bag and placed them on a hand-painted board.

Kevin looked at the ragtag setup and thought, *This'll be easy.* He could already beat his uncle at chess. Kevin stuck out in his own family as unusually precocious, one of those rare kids who never showed any natural fear of adults; ever since he was little, he had spoken to them as if they were his equals. He had a talent for talking, and when he was five, his parents had taken to calling him Little Old Man.

Paul moved his pawn two spaces forward. Then he moved his knight and his bishop in a pattern known to chess enthusiasts as the Giuoco Piano opening. Kevin didn't yet understand the concept of a standard chess opening, its mathematical power and elegance, its way of imposing structure on a nearly infinite game. Paul checkmated Kevin in only a few moves.

Stunned, Little Old Man did something he had never done before: He apologized to an adult.

"No, no," Paul said, "you're about where I thought you would be." He was being kind; Kevin found out later that Paul had a master's ranking.

Most other thirteen-year-old kids would have blown Paul off, but Kevin was different. He didn't like school. His classes bored him, and he got mediocre grades. But there was a hunger in him. If a master beat you at something, the proper reaction was not irritation, but humility. You didn't get mad; you said, *Teach me what you know.*

After that first defeat, Kevin and Paul ended up playing every weekend for a year. As Kevin's skills improved, their games stretched out for hours, and they would talk while they pondered various moves. Kevin had no idea how Paul knew all the things he knew about—architecture, physics, computer programming—but it didn't matter *how* he knew, only that Paul was the rare adult who took Kevin seriously, who found the kid's energy charming instead of weird. The two of them would go to used-book sales at the Park Forest library and return with paper grocery bags overflowing with books about art and science. Paul would point out pictures of student building projects—toothpick bridges, robot battles to the death. He would talk about past winners of the Nobel Prize in Physics and what they had discovered.

Once, when Kevin was fifteen, Paul showed him a picture of a group of men in gray suits and bowler hats staring at a massive concrete pillar that seemed to be upside down: skinny at the bottom, fat at the top. It had been designed by Frank Lloyd Wright for use in the future headquarters of the Johnson Wax company in Racine, Wisconsin. Wright believed that his "mushroom pillar," reinforced with steel mesh instead of traditional steel rods, would hold greater amounts of weight than conventional designs, but the state's industrial commission didn't believe him. So Wright arranged a test. He built a platform atop the pillar and piled it with sand. A June 4, 1937, article in the *Milwaukee Journal* described the scene:

> *At 4 p.m., after 18 tons of sand had failed to crack the pillar, workmen and visitors retired to the company recreation building for beer and pretzels, a breathing spell, and a short talk by the architect. After the respite, workers, with the aid of a derrick, continued the task of distributing the weight of sand evenly across the wide top of the steel mesh and concrete post. At 6 p.m. the structure was still standing, and plans were made for continuing the test Friday, adding weight until the column crashes.*

Kevin loved that. *Here, try to break this, I dare you.* And the doubters couldn't break it, not even with sixty tons of sand.

More and more, Paul's comments struck Kevin as calculated,

purposeful. By now, 1986, Kevin had seen the *Star Wars* trilogy, and it was as if he had found his own Yoda, right here in Park Forest—a mysterious person who would reveal secret knowledge only to those worthy of possessing it. Yoda seemed to be laying down the stones of a future path. He had evidently decided that Kevin was ready to learn about the Force. The Force was called engineering.

Engineering, Kevin learned, was not the same as science. Science was a method for discovering things you didn't already know; engineering was about making stuff *with what you already knew*. And that stuff could be grand, like a Frank Lloyd Wright building, or humble, like a toothpick bridge. One year Kevin mentioned that his high school was sponsoring a toothpick bridge–building competition, and Paul decided that Kevin was going to win. He talked to Kevin about trusses, the shapes composed of triangles that give bridges and other structures their strength. Other shapes can be distorted if pressure is applied—a square, for instance, can be tilted into a rhombus—but for a triangle to change shape, it has to fail at one of its three vertices. The whole thing has to break at once.

Kevin realized how easy it was to make a superior toothpick bridge. First you sift through the boxes of toothpicks to find the ones without any veins or cracks. Then you lay them out in triangles on wax paper. The triangles are trusses. Squirt the glue, let it dry. Add weights to the bridge until it snaps; then tweak the design, build another bridge, add more weights until the new bridge snaps. Kevin's final bridge design used forty-three toothpicks and weighed less than a twentieth of an ounce, about the same as a single leaf of lettuce. It held 51 pounds of weights before snapping. The second-place bridge held less than 10 pounds.

Something about the challenge of reaching the ultimate failure point of the materials appealed to Kevin on a deep level. The toothpick bridge was his first taste of destiny, his first sensation of doing what he was meant to do. The world is made of stuff and that stuff behaves in certain predictable ways; if you master the rules, you master the stuff. "Paul showed me these things for a reason," Kevin would later say. "He believed in me, that I could do it."

Light a man on fire and he'll be warm for the rest of his life.

In 1990, when Kevin was getting ready to leave Park Forest for his freshman year at the University of Illinois–Chicago, Paul said something a bit unusual. Casually (though nothing with Paul was truly casual), Paul told Kevin that he was certain that within twenty years, everyone in America would be driving some sort of solar or electric car.

Yoda smiled and nodded at Little Old Man.

The message had been delivered.

I guess that's what I'm building next, Kevin thought. *Some sort of solar or electric car.*

8

Wingless

n 2004, an electrical engineer in Carlsbad, California, started building a car in his garage. His name was Steve Fambro. He was thirty-six years old. At work, he made robots for a San Diego genetics company, but he was bored. He'd been tinkering with cars since he was a kid, and now he was thinking about how gas prices were rising and oil was finite and the world was warming and yet no auto companies were addressing these realities in a bold way. Surely he could do better. He could design a small, nimble, inexpensive vehicle on first principles—lightness, aerodynamics, safety—and he could produce it and sell it to people like himself, people who had long sensed that a vastly more efficient car was possible.

Soon Steve was spending all his spare time on the car. Then he quit his job so he could work on it uninterrupted. "When you tell people you want to start a car company," he told the local newspaper in 2006, "you might as well be telling them you're planning on going to the moon. My wife complains she is a car widow." But he made fast progress. Various California characters attached themselves to the project: first a physicist who helped him model the car's aerodynamics in a computer wind tunnel, then a guy who made wake boats. The boat builder, Chris Anthony, knew how to create arcing curves of strong, light, aerospace-composite material, exactly what you'd need to build a car shaped like a teardrop, one of the most aerodynamic forms known to man. Another way to visualize the shape was to think of a wingless bird or, as the two guys would later explain on their website, "a linebacker poised for the blitz." And because this was California, where multibillion-dollar companies have been founded on less, Steve and Chris decided to go into business together.

They called the company Aptera Motors, *aptera* being Greek for "wingless." They made a series of prototypes and showed them to journalists. The prototypes were powered by battery packs and electric motors, and they had three wheels instead of four, to save weight: two in front, one in back. The front wheels extended away from the body and were shrouded to minimize wind resistance. With their smooth, seamless, iPod-white composite bodies, the Aptera cars looked like sculptures you'd see in chic Danish apartments. These were clearly California vehicles, imbued with the optimism and impertinence of the technology industry. Silicon Valley thinking had revolutionized so many American industries—finance, retail, publishing, entertainment. Why not autos, too?

In 2007 and early 2008, green-energy sites and tech magazines wrote dozens of upbeat articles about Steve Fambro and the Typ-1, as he was calling the vehicle then. Aptera announced that anyone who wanted to buy a vehicle could reserve one for $500, and eventually more than four thousand people put down deposits, among them, reportedly, comedian Robin Williams and basketball star Shaquille O'Neal. Google's philanthropic arm invested $2.75 million, and Steve raised $20 million on top of that. The company grew. It seemed to get closer and closer to making cars. In early 2008, it announced it would begin selling vehicles in California by the end of the year, priced at between $25,000 and $45,000 each.

Later in 2008, Aptera moved from Carlsbad to a 75,000-square-foot production plant in Vista, California, a 45-minute drive north of San Diego. Employees scribbled inspirational messages on a white wall:

Apteraction!
The Art of Belief
I believe, lightweight composites taste good!!!
All things are possible to him who believes.—Mark 23

By mid-2008, though, the nation's fortunes had begun to tip. The stock market was sliding into the tar pit. The economy was deleting mighty investment banks as casually as spam e-mails. Aptera needed another major infusion of capital to get its car into production, and money was suddenly difficult to come by.

Like so many other struggling companies that year, Aptera turned to the U.S. government. In December 2008, it applied for a loan from the $25 billion Advanced Technology Vehicles Manufacturing Loan Program, or ATVM. Administered by the Department of Energy, the program had been signed into law under President George W. Bush, but in terms of money dispensed, it was a mere creek until Barack Obama came along and turned it into a river. The basic idea was that since the private sector had failed to invest in fuel-efficient vehicles, the government should step in and help companies get their products off the ground. During the Obama administration, the program would loan $5.9 billion to Ford to increase fuel efficiency on thirteen of its models, $1.6 billion to Nissan to help it bring its Leaf electric vehicle to market, $465 million to Tesla Motors, and almost $200 million to Fisker Automotive, a California company owned by a flamboyant Danish auto designer who wanted to manufacture a luxury hybrid sport sedan in Finland.

Aptera asked the government for $184 million. The DOE quickly rejected the application: Aptera's car had three wheels, and the regulations allowed funding only for cars with four wheels. Meanwhile, something dramatic began to happen back in California.

The car that Steve Fambro envisioned had always been an uncompromising one. For instance, to maintain aerodynamic efficiency, he'd made it so that the windows couldn't roll down, which meant that if you wanted to get a burger at a drive-through, you had to open the whole door. But Steve also realized he needed help to move beyond the prototype stage. Recently, he'd hired several people from Detroit—people with experience in getting cars into production. He'd even given up the role of CEO, assuming the title of chief technical officer. The new CEO, Paul Wilbur, had spent decades in Detroit, serving in high-level roles at Chrysler and Ford.

Now, though, Paul Wilbur was trying to make major changes to the car's *design*. As the company wrote in a letter to depositors in January 2009:

> *For months we have been receiving important feedback from you, our depositor community, and we have come to realize there were flaws in our initial product assumptions—specifically as it pertains to satisfying the needs of real-world consumers. Our greatest degree of learning came just a few months ago when we asked all of you to participate in a brief survey. This critical piece of research requested insights about your expectations for our company and our products, and we discovered a notable disconnect between our product plan and realistic expectations. Some modifications had to be made. For example, you helped us realize that some trade-offs for convenience (like being able to grab a burger in a drive-thru) might be necessary to make the ownership experience more palatable, even if it cost us a couple tenths of a point on our drag coefficient.*

At this point, the car had already been in development for five years. The new changes meant it would be delayed even further.

Later in 2009, Steve decided he'd had enough. He tried to oust Paul from the board. It backfired. The board supported the guy from Detroit instead of the guy from the garage. Steve and Chris were booted from the company they'd founded.

When the news got out—APTERA FOUNDERS OUSTED IN BOARD-ROOM SHOWDOWN, read the headline on Wired.com—many of Aptera's most zealous fans were livid. They'd supported the company because it represented everything Detroit was against, and now here were these Detroit guys messing everything up. They didn't want the car to be perfect, they just wanted it to be done; the company could improve it on the fly, in subsequent versions, the way software companies do. "Classic case of assuming that 'experts' in a failing industry know anything worth knowing," wrote one Internet commenter. "They should have stuck with their vision and not tried to follow Detroit into the dustbin of history. Anyone who is worried about opening windows, go buy a Ford."

Aptera's situation deteriorated over the next few months. In November 2009, it laid off ten employees.

With the help of friendly California legislators, Aptera successfully lobbied to change the ATVM rules to allow funding for three-wheeled cars, then resubmitted its application. Company executives said they were back on track to go into production. But no one believed them. The company's history of rosy statements and repeated delays had drained Aptera of all credibility. It was now stuck. It couldn't go into production without investment, and it couldn't attract investment without convincing people it had a car worth producing and a means to produce it.

If Aptera was going to survive, it needed a way to restore its credibility. It needed a trusted third party to come in and vouch for the car—to say that it was real, that it could do what Aptera had always said it could do, and that it was worth waiting for. And that's why, in early 2010, the future of Aptera and its twenty-five employees became inextricably tied to the X Prize.

———

CHIEF MARKETING officer Marques McCammon hands me a rubber sledgehammer.

"Hit it," he says. "As hard as you want."

It's a balmy day in April 2010, a week and a half before Shakedown, and we're at Aptera headquarters in Vista, in the spacious warehouse the company calls its production plant. Marques is standing in front of one of several prototype cars. I glance at him.

Really?

He nods. I swing at the car. The hammer connects with a dull *gong*, sending a shock wave up my arm. It leaves no dent.

"If you do that to any steel-body vehicle, you at least dent it, at most put a hole in it," Marques says. The car's body is made of glass-based fabric and resin surrounding a honeycomb core. "Our composite structure actually dissipates the load. It acts like a million little I beams."

Marques is thirty-three, with a chipper demeanor and the karate-stance readiness of a seasoned communicator. Before coming here, he worked at both Chrysler and Saleen, a specialty automaker. *Real Detroit Weekly* once wrote an admiring story about Marques's reputation for "designing new rides and understanding the market," saying that it was "only a matter of time" before he was running his own car company. He's a hybrid: an entrepreneurial striver who left Detroit for a reason but still appreciates what Detroit does well.

"I'm not one to defend Detroit," he tells me, "but they have to make sure weld one is the same as weld five hundred. I talk about this to my depositors. They say, you already have a body. Well, no. They say, You're just trying to make it more expensive, drag it out, make this like Detroit. *No.* Listen, I could write a book about what's wrong with Detroit, *believe me*, but this is an industry that's been around for a hundred years, survived every major

The Aptera 2e.

economic catastrophe, and grew three and a half percent a year. They're doing *something* right."

Talking to Marques, I get the sense of a guy dealing as best he can with a couple of communications challenges so unusual they border on the surreal. The main one is that his customers are *too enthusiastic* about the car. They want it now, and he has to explain to them, again and again, why they can't have it yet. "I had a lady who e-mailed me and said, 'What are you guys doing?'" he says. "'Hurry up and get the vehicle to me. The only thing I care about is that you give me an affordable three-wheeled vehicle that's reliable, that drives in hot and cold and wet weather, and I can depend on it to drive me to work every day.' Okay. So let me tell you what it takes to get there." It's one thing to make a prototype, McCammon says: You mix resin in a large bucket, you pour it into a form, you let it harden, then you mix a different bucket of resin to make a second prototype. It's exponentially harder to make a thousand cars so that the resin in the thousandth car is chemically exactly the same as the resin in the first car—not to mention making sure those thousand cars are comfortable and reliable and meet federal crash-test standards. (As Tesla cofounder Martin Eberhard has pointed out, "A car crash is fundamentally different than a software crash." In a car crash, you can die.) This is what Aptera is struggling with right now, as the company develops the latest version of what it now calls the 2e: the car it intends to take into production, the car it will compete with in the X Prize.

The fact that the production-intent car and the X Prize car are one and the same is Marques's other big challenge. Normally, car companies test their cars in private, so that no journalists will see the inevitable bugs, mechanical failures, and crashes. Aptera doesn't have that luxury anymore. It needs the Prize to prove that its car has virtue, and the Prize is open to the press. Competing, then, is a huge risk, but Aptera has no choice. If the car breaks down, Marques will have to spin apparent failure as the price of success—an inevitable stage in the messy process of developing a car.

Right now, the 2e is hidden behind a high wall in the warehouse. I hear sounds of drilling and cutting. Marques says he can't let me peek. The guys "are heavy into race prep," he says. "I apologize, but if I asked them to stop I'd be risking my life."

THE NEXT day, Aptera holds a press conference at an airplane hangar in nearby Palomar. Inside the hangar, employees circulate in identical black shorts and black-and-gray racing jerseys. There's a makeshift stage, a patch of concrete in front of a long black curtain. Ten circular tables flank the stage, decorated with white cloths and crystal decanters holding white daffodils. Green uplights make the curtain glow. A DJ with black glasses and a goatee twists knobs in a booth. In folding seats, several dozen reporters, Aptera sponsors, and bloggers—men in their twenties with beards and digital cameras around their necks—await the start of the event.

The lights go down. The DJ cues a U2 song. In the darkness the noise swells, then subsides. The DJ says, "And now, the president and CEO of Aptera, Paul Wilbur." Paul, wirelessly miked, wearing a mud-colored blazer, steps to the center of the stage.

"Thank you all for coming out to a somewhat unusual place for an automotive press conference," Paul says. "Although an airplane hangar honestly isn't that strange to Aptera." He brightens. "I've been telling anybody who will listen to me that the Aptera flies right by gas stations for about a year." He flattens his right hand, lifts it to head level, and makes it zoom through two feet of atmosphere. The gesture is endearingly kid-like. Bloggers touch-type on laptops, check Facebook. Paul blinks, turns, and walks to the other side of the stage.

"So," he continues, "you may have noticed recently that we've kinda been a little quiet the last few months. You know, I have a journalist friend in Detroit who told me, 'Hey, Paul, you better shut up until you have the hardware to back it up.' And I know what all of you are thinking"—Paul grins—"and actually I *do* have a journalist friend."

After an uncomfortable silence, Paul continues, "No one ever said that building a vehicle was easy. There have been some really unavoidable product and financial hiccups . . . so yes, we are behind, we are behind . . . but recently we've turned a very important corner. . . . Mr. Reichenbach?"

The curtain ripples, the DJ cues a bouncy beat, and Tom Reichenbach, the company's chief engineer, drives the new 2e through the curtain, flashers blinking. Compared to the prototypes, it looks bigger, wider, fuller, but otherwise the distinguishing features are the same: three wheels, avian profile, white composite body. When Tom climbs out, Paul introduces him as a former leader of the GT racing program at Ford and a champion sports-car driver.

"Paul asked me once why I came to Aptera," Tom says. "I

said ... competing this vehicle on the track at the X Prize will be a nice balance to the other cars that I've done. They're all gas-guzzlers. So if I want to get to heaven, I have to do something responsible."

Tom walks around to the back of the car. He demonstrates that the trunk includes enough space for two sets of golf clubs and a gym bag. The doors are bigger now, and the car is easier to get into and out of. He says the windows roll down now, then rolls them down to prove it.

After the show, Tom and Paul invite people to examine the car up close. Tom stands at a distance, watching depositors and journalists peek through the windows and into the trunk. He says he's looking forward to the start of the Prize. "It's *exactly* like going to a race," he says, smiling. "A lot of our guys here have never been to a race!"

I ask him if he's ever heard of Edison2. He nods. "They're racers, so when they show up, it'll be game on," he says. Beyond that, though, "We don't quite know what's going on there."

9

The Hypocrisy the
Masturbating Hand Endures

The race mechanic Bobby Mouzayck had never met Oliver Kuttner when he got a call from him in late 2009 with the offer of a job at Edison2. On the phone, Oliver sounded like "this mad-scientist lunatic," Bobby recalls. "Just totally like: 'Yeah, whatever. Get a ticket, come on out. You wanna come out, like, tomorrow?'"

Oliver needed a good mechanic. He wanted Bobby to move to Lynchburg and help his team assemble the Very Light Cars. To make it easier for him to move, Oliver said he'd let Bobby live rent-free in a studio apartment in back of the shop. He'd lend him some furniture to make it cozy. If Edison2 won the X Prize, Bobby would get a bonus out of the prize pot. And afterward, depending on the company's success in attracting investment, maybe there'd be some kind of a permanent job.

It sounded good to Bobby. He'd been working on race teams for several years, bouncing from Utah to Florida to Colorado, where he was living now. But his current team was running out of money. Bobby sensed that it was only a matter of time before it folded, putting him out of work.

What the hell? he figured.

A few weeks later, he arrived in Lynchburg.

IT ALL came from his father: his knowledge of cars, his love of racing, everything. Bobby was only four or five when his father started teaching him. Robert Mouzayck owned an automotive heating and cooling shop in Atlanta. Bobby worked in the shop from a young age. The shop was prosperous. Robert drove multiple Cadillacs, many of dubious origin, including a two-tone ride with a picture of a naked woman on the hood.

Robert had lots of colorful friends, including one who worked on Paul Newman's racing team. After one race, Bobby was allowed to sit in the cockpit of Newman's white-and-green Lotus Esprit. *God, look at this weapon*, Bobby remembers thinking. Another time, when Bobby was eight, his father took him to his first drag race. Somehow, they got separated in the crowd of thousands, and when Bobby told a security guard he was lost, the guard took him up to the top of the control tower so the race officials could page his father. As soon as Bobby walked into the room, he saw he was inside the strip's luxury suites, with a clear, godlike view of the two drag-strip lanes and the hot rods trailing fireballs. An attractive spokesmodel for one of the race's sponsors offered Bobby an ice-cold Coke. When Robert came to collect his son, he looked around at the view and the model. "Shit," he told Bobby, "you should get lost more often."

As intrigued as he was by racing, Bobby didn't set out to make it his career. In 2000, after graduating from high school, he enrolled in a small private art school in Burbank, majoring in architecture. He wanted to be either an architect or an engineer. For a while, college seemed to work out. Bobby joined the rugby team and developed a reputation as a fearsome presence in a scrum. He never had any trouble finding girlfriends. Tall and athletically built, with dark, curly hair, Bobby came across as both boyish and slightly dangerous, thanks in part to a diagonal scar that connected his right nostril with his upper lip—the remnant of a childhood accident. He liked his architecture classes. Good with his hands, he produced high-quality drawings and models. He took pride in the fact that the work he presented during class critiques was neat, precise, structurally correct, and, above all, plausible. His designs were ready to be scaled up to something bigger in the real world.

But when his professors looked at his stuff, they shrugged and moved on. They gave him C's, lavishing praise—and A's—on the more abstract, speculative work of his classmates. This drove Bobby crazy.

One day, after a professor told him that his work was "lacking in emotion," Bobby decided to perform a little experiment. For his next assignment, he would submit a project so ridiculous, so over-the-top emotional, the professor would have to recognize it as a fraud. The assignment was to build an eight-inch wooden cube containing a space for a human hand. The hand, once inserted into the cube, could be making any number of gestures, like a peace sign or a thumbs-up. Bobby executed his cube with all the skill he could muster. Every joint was tight, every edge level. Here's the description Bobby submitted with his project:

The hand I chose to depict is one that is masturbating. Because of the hand's actions it should be concealed. Thus I made a cube with a door that could isolate it from the outside. During the planning stages of this project one asked, "Is the hand's

action sexual or rather more of a social escape?" Shunned by
the outside world, yet praised by the inside is the hypocrisy the
masturbating hand endures.

This was the only A Bobby received in architecture school. His professor told him he was finally expanding his mind, and his classmates gushed. "They were jerkin' off all over it," Bobby remembers. By praising his bogus project, they had failed his test. "That's when I knew architecture wasn't for me."

Bobby started thinking about his $25,000-a-year tuition. It wasn't fair to make his parents pay that, especially since his father's auto business had fallen on hard times. Robert Mouzayck, depressed and increasingly desperate, had begun to rack up large gambling debts. Bobby left California for Alabama, where he enrolled at Auburn University and switched his major to mechanical engineering, but after the regents approved a 40 percent tuition hike, Bobby decided to leave college and join a race team in Colorado to make some money.

As soon as he began racing, he realized he was in the right place. In school, he'd enjoyed looking at cool buildings. But on the last lap of a race, when the drivers were fighting for every inch, Bobby felt something in his skin. When he was working on a car, and the pressure was on, he didn't need to eat or sleep, and he didn't understand why other people needed such things. Something happened to his body—a slowing of the breath, a relaxation of the muscles. His work ethic caught the eye of Ron Mathis, who worked with Bobby in Florida, on Audi's race team. The engineer was impressed with Bobby's "ability to filter out stuff around him when he really needed to concentrate," Ron says. "He's a tremendously physical guy."

In Florida, Bobby's job was to take orders. Race teams are top-down organizations. The guy who owned the team was an old friend of Oliver's named Dave Maraj. Everyone called him the

King. According to Bobby, the King's philosophy went like this: "You better look good, and your shirt better be tucked in, the floor better be painted white, and it better be done on Friday like you said." Bobby responded well to that. *Don't give the King any shit. Just give him what he wants. And if he wants shit, give it to him.* But that didn't mean he could shut off his brain. Ordered to perform some task on a car, he wouldn't simply execute the task. He would think about how the part he was working on fit into the whole and whether the fit was optimal and, if it wasn't, how the whole car could be more rationally designed. Then he'd tell the engineers what he thought, drawing on what he'd learned in his college classes: "I'll *do* the work as a mechanic, but I'm gonna *tell* you as an engineer: 'This shit is fucked, you're out to lunch.'"

THE FIRST thing Bobby noticed when he started working at Edison2 in 2009 was the company's lack of manpower. In racing, there are usually multiple mechanics for each car, but Edison2 only had three main mechanics, including him, for four cars.

The lead mechanic was Peter "P.K." Kaczmar, a fifty-four-year-old native of north London. P.K. worked on the main shop floor, the big squarish room, next to an immaculate tool chest and bench adorned with tools arranged at perfect right angles. A spray bottle bore the label P.K. DON'T REMOVE. Acerbic and solitary, with a rim of light gray hair, P.K. had left school at age sixteen to learn the trade, but his mind was sharp and he read voraciously; once, in the shop, at the end of a low-key argument about some aspect of the car, P.K. smiled and said, "What was it Oscar Wilde said on his deathbed? Either these drapes go, or I do." He subscribed to *Skeptic* magazine and seemed to view himself as Edison2's in-house curmudgeon, a necessary counterweight to the optimism of Oliver and the engineers. He would sometimes antagonize Ron by taking a power drill over to his desk and casually ramming a drill bit as roughly as possible through some gorgeously machined piece of aluminum.

The other mechanic was Reg Schmeiss, a native of Hahndorf, Australia, an hour from Adelaide. Reg had dropped out of school at fifteen. Since then, he'd traveled all around the world, building and maintaining cars. He spoke four languages. He was married to a French woman and wore his brown hair in a mullet. Bobby and Reg got along great. They worked together in a part of the office invisible from main shop floor, down a wide hallway lined with vintage racing posters, the cherry-red chassis of a rare Bizzarrini sports car, and the wrecked cars Oliver salvaged from a crash-testing facility. This was the body shop, an ecosystem all its own. The only source of light was a series of ten individual bulbs hung from the ceiling on ropes. A yellow Pirelli banner read: POWER IS NOTHING WITHOUT CONTROL.

Bobby's first big task at Edison2 was to get a grip on the composites situation. In the long run, Oliver wanted the car to be made almost entirely of aluminum and steel, to reduce cost and boost environmental friendliness. But his more immediate goal was to win the X Prize. Millions of dollars were on the line. So, for now, Oliver and his engineers had decided to veer away from Oliver's high-minded ideal. The bodies of three of the four Very Light Cars would be made of carbon-fiber composite instead of the heavier fiberglass used on the fourth. To create the carbon body panels, specialists at another company somewhere would begin with cloth woven from carbon "yarn," each strand composed of thousands of individual carbon filaments; the technicians would lay the carbon cloth into a mold, then add a resin, then suck the cloth tight against the mold using a vacuum bag. And the cars' wheels would contain a mix of magnesium, aluminum, and carbon fiber. Each wheel would weigh a scant 6 pounds. (Edison2 didn't have a choice but to create a custom wheel; no existing wheel fit the car's needs. The Very Light Car had to have narrow tires, to minimize "rolling resistance," the loss of energy due to friction with the road, and a large wheel diameter, to make room for the custom in-wheel suspension. Ron says he "makes no apologies" for using expensive materials in the Very Light Car's wheel: "Those who don't go the extra mile, lose.") By weight, about 60 percent of the final car would be aluminum, 20 percent would be steel, 10 percent would be carbon, 2 percent would be magnesium, and the rest would be stuff like copper wiring and seat padding.

Edison2 was farming out the carbon fiber work to a shop in Ohio. When the body panels arrived in Lynchburg, they looked like the curved fragments of eggshells laid by some enormous bird: thin, smooth, fiercely sculptural. The panels had to be trimmed and sanded and shaped to fit atop the car's steel frame. Although Bobby wasn't an expert at this kind of work, "nobody really wanted to do it," he says, "and the shit needed to get done, so I just started doing it." He'd spend hours shaping the corner of a particular form so it matched, as identically as possible, the corresponding forms on the other cars. You had to really enjoy working with your hands to do it well, and Bobby did, but he got sick of sanding all day. Dark-gray carbon dust filled the air and coated every surface—when Bobby walked back to his apartment to go to the bathroom, he tracked it into the bathroom on his shoes—and he could feel himself breathing it in, a toxic mist of stronger-than-steel filaments. "It's, you know, cancer," he says. (He sometimes used a mask, but other times he was too busy to spend five minutes digging for a mask underneath the body shop's pile of rubble, and he'd just pull his shirt up over his nose and get it done.)

It was dirty work but not brainless work. Every aspect of the car had been measured and mapped in Ron Mathis's CAD software, so, in theory, all the parts should have snapped together neatly. But building a car from scratch was turning out to be more complex than that. "It's not all blue screens and high fives," Bobby says. "That's not the way it works."

A big problem was that X Prize officials were constantly changing the rules, making seemingly minor tweaks to the format of
competition events and the regulations governing the vehicles. The
officials had reserved the right to change the rules as they went, but
it drove the men of Edison2 crazy. They were racers, taught to sift
the fine print for any possible advantage. No race worth a damn
would ever change the rules in midstream. But the Prize wasn't a
race. Sometimes the officials made some minor-seeming change
that turned out to be not so minor in the implementation, or suddenly clarified a requirement that had been vague, and either way it
meant more work for Ron, therefore more work for Bobby and Reg.
Or sometimes, when parts arrived in the shop, they'd be wrong,
and the mechanics had to figure out how to modify the part to fit
the car or the car to fit the part. Like the wheel-pod debacle. The
wheel pods were supposed to be reinforced with extra material on
the side that faced out, to provide additional protection in a crash.
But when Reg received the wheel pods, they'd been reinforced on
the wrong side. So he was asked to just build a reinforcement on the
other side. "'Just' is their favorite word," Reg says of the engineers.
"Can you just do this? 'Just' can lead into twelve hours, you know?"

These projects often required hours of thought and trial-and-
error handiwork. Without consulting the engineers, the mechanics would whip up little drawings and machine parts themselves,
or else have Edison2's in-house machinist make the parts. Bobby
often lay awake at night and thought about how to make the car
better. Everyday objects—the curved slat of a venetian blind, the
rounded end of a window ledge—would spark ideas about wheel
pods or fairings. "You're never off the clock," he says. "If you are
true and passionate in your trade, whatever your trade—I don't
care if it's blow jobs or nuclear bombs—you never stop thinking
about it."

The more Bobby obsessed about the car, and the more auton-
omy he was given, the more he started to think of himself as not

just the car's mechanic, but its cocreator. After all, Edison2 wasn't a traditional race team with a clear leadership hierarchy and rigidly defined roles. And it wasn't like they were making a laptop or a children's toy, a product designed by educated people in the U.S. and assembled by unskilled workers in Asia. The Very Light Car was more like inventions of old: a thing built by the hands of its imaginers, the final object formed by many small improvisations. It was an ever-evolving beast shaped to meet the needs of a competition with ever-changing rules—the ultimate instant adaptation.

The engineers didn't necessarily see it this way. Bobby learned this when he tried to tell them about the problems he saw with the car, which ranged from the radiator design (too many bleeds and auxiliary hoses) to the wheel pods (which would make it hard to change a flat tire) but always boiled down to the same core concern: The car was too complex for the time available to test it. The engineers, in his opinion, had tried to make everything too perfect, too beautiful. In Reg's phrase, they were "building a monument to themselves." But perfection might come at a cost. No one knew how the car would perform at the track; the engineers were making educated guesses, Bobby thought, and a guess can be wrong.

"It's better to have half the car and test it twice as long than have twice the car that you never fuckin' tested," he says. "You think you're that great? You think it's just gonna *work*? No. No no no no no no no. Especially with this project, where we have to push it *so* far."

But the engineers had confidence in their judgment, and Oliver generally deferred to them. When Bobby complained to Ron about something, Ron tended to listen respectfully, then keep doing whatever he'd been doing. (Asked about the relationship between Edison2's engineers and mechanics, Ron explains that most arguments boil down to this: "The engineer says, 'Well, it isn't that bloody easy when you just gotta make this thing out of thin air.' And the race mechanic says, 'Yeah, well, you should have done a better job.'

Most of the time, the mechanics are right. You see some really bad stuff appear. But both sides keep each other honest.") As for Barnaby, he didn't seem to listen at all. When he visited the shop, as he did a few times, flying in from California, he filled the place with a jittery, emphatic energy, speaking mainly in pronouncements. The mechanics started to dread the sight of Barnaby's white beard. Behind his back, they called him Father Christmas.

Still, as much as the decision-making process at Edison2 frustrated Bobby, he enjoyed his job. The car was interesting to him. It wasn't as good as he wanted it to be, but it had real purpose and meaning. Maybe the first version wouldn't reach 100 MPGe, but Bobby would take 70. And then they could improve it. Maybe, by version five, you'd get in the back, with the A/C on and the stereo cranking and the car doing a legit 100 MPGe—"That's huge! That's a hell of a success. I do think there's a future for it, for sure. And I would love to have, especially in this economy, a secure job. That would be great."

Bobby also liked working for Oliver. He respected his boss's unusual style of leadership. Instead of ruling by fear, Oliver conveyed an intense sincerity that went a long way. "He's the kind of guy who will cry because he's so proud of you," Bobby says. "And I firmly believe that when a grown man cries, he's not fuckin' lying." People were always coming into the shop needing to conduct some transaction or another related to Oliver's many businesses, and Bobby noticed that Oliver treated everyone with the same respect, whether they were his investors or his low-income tenants, guys with vague mustaches and soiled hoodies who would ask, with downcast eyes, if *Mistah Kuttnah* had a minute to spare. Oliver's most crucial skill was a simple ability to talk to people and make them feel welcome in his presence. And this often translated into cash. "Oliver's ability to find money and woo people into his business is absolutely the best I've ever seen," Bobby says.

Once, Oliver came to Bobby and said, "Listen, I can't pay you this week. I know I already owe you from last week. However, last month I was able to find three hundred grand just by talking to my friends. Have faith in me." Bobby told him it was no problem. "I never doubted him," Bobby says.

OLIVER WALKS out of a Charlottesville coffee shop and into a spitting rain. It's a weekday morning in April, five days before the start of Shakedown. He's wearing jeans and a dark blue T-shirt full of holes, spattered all over with white paint. His gray hair is an experiment in static electricity. He hasn't shaved.

A middle-aged black man approaches, mumbling something about a broken window. Oliver pulls out his wallet and hands the man a crumpled $50 bill. "Thanks, Frankie, you're the best," Oliver says. Frankie walks away and Oliver shrugs theatrically. "Who knows where it goes?" he says.

He explains that Frankie lives for free in one of his buildings, in exchange for doing small repairs. "I take care of him, and he takes care of me."

We climb into Oliver's pickup, bound for the Edison2 workshop in Lynchburg. On the armrest are two packs of TNT smoke balls, low-intensity fireworks—little gifts for his kids, purchased out of guilt. Last night, he says, "my wife was very upset about how many hours I'm spending on the car right now. I'm giving the whole burden of raising the four kids on her lap."

As we drive through town, he points out the Terraces, a single building chopped up to look like many different ones. It has no roof, only a Tetris-block pattern of stepped terraces—a trick designed to preserve the illusion of organic individuality and natural growth. "It's the youngest building in the state of Virginia to be designated a historic landmark," Oliver says. "You can't tell, but it's the largest building in downtown. That's a building that I built with a bunch of guys."

The rain intensifies, crashing violently against the truck's windshield. Oliver pulls into the parking lot of a local bank. He grabs a manila folder and sprints to the door. Ten minutes later, he sprints back and hands me a stapled packet of paper. "Bankers want to lend money to people who don't need it," he says, beads of rain dripping down his forehead. "So you can't look like you need money. Okay? That's just the way it is." I look at the top sheet of paper. It's a signed commitment letter for a $1.3 million loan.

So far, Oliver has been funding Edison2 with money from small investors. Most are wealthy men in the Charlottesville area who've done deals with him in the past. The first to invest, in 2008, was a guy named Larry deNeveu. I went to visit deNeveu not long ago at his home in Free Union, outside Charlottesville. A reedy, sandy-haired retiree, deNeveu used to run his own ship-brokering company, an ancient trade that involves matching cargoes with ships. Sitting at his kitchen table, near a sign that said THE DOGS FROM HELL LIVE HERE and a well-thumbed copy of *The God Delusion*, deNeveu told me he'd long been interested in efficient cars—the Smart car in the driveway was his—and he liked what Oliver and Ron Mathis had to say when they first came to his house to pitch him. They told him they'd read the rules of the X Prize and had figured out how to win it; they told him about lightness, and argued that a very light car with a small gasoline engine could beat electric cars because the gasoline engine is tried and true and electric cars are still buggy. "Our car starts every time you push the button," Oliver said. "If you want to race, first you must get to the race." Oliver and Ron would make a very light car to win the X Prize, and in the process, they'd invent technologies they could license to automakers, paying back investors with the proceeds.

DeNeveu asked Oliver why Ford or some other big automaker hadn't released a very light car. He was then "anointed with the gospel from Oliver," which went like this: The engineer from Ford gets a paycheck from his boss. And he's not about to tell his boss that all those chassis parts, and all that tubing they have on order, and all those rims, and, oh yeah, that *factory* over there, with that assembly line they use—that's gotta go. Because his job will vanish the day he speaks like that. Oliver told deNeveu that even if Ford wanted to build radically different cars, it simply couldn't. "In a way," Oliver told deNeveu, "it's a luxury to start with nothing." After half an hour, deNeveu stood up, said, "Let me go get my checkbook," and sent Oliver away with a check for $50,000, also committing to give Oliver a few hundred thousand more as he needed the money.

DeNeveu told me he decided to invest in Edison2 because he was impressed by the logic of a light, gas-burning car: "This is about looking at the results and being rational and embracing reason." (As soon as the first chassis was built, Oliver sent deNeveu a picture of two of his young children lifting the car and grinning.) But maybe more important, deNeveu just really liked Oliver. He'd lent money to Oliver before, and not only had Oliver always paid him back ("I know he's not going to tell me to pound sand up my ass"), he'd also thrown his whole heart into his projects, working hard enough to risk domestic pain. "I know it annoys the hell out of his wife," deNeveu said. "She's a younger person, and she is tired of this crap, him not being with his family." If Oliver said he was going to win the X Prize, deNeveu knew he would uproot his whole life to get it done—although deNeveu did worry that the officials might tilt the rules against Edison2.

Later, Oliver seemed to confirm this fear when he told deNeveu that the Foundation had switched formats. Originally a cross-country race, the Prize would now be held at a NASCAR track. Also, the $10 million pot, long promised to a single winning team, would be split three ways. "I thought, *What?*" deNeveu recalled. "That's like sitting down to play Monopoly and somebody hands you a roulette wheel. I don't get to pass Go? These are not the rules that come with the box! And that's when Oliver had the mind-set: 'I don't care what they do. You could put nine categories out there. I'll build nine different cars.'"

After deNeveu cut his check, others followed suit. There was Dr. Wolf Zweifler, aka "Wolfie," a German-born attorney and blue-jeans dealer who had met Oliver years ago at a car auction in Arizona. "He was a classic-car dealer," Wolfie told me. "Classic-car dealers do not have a very good reputation. He didn't always fulfill his prognosticating, but you came out well in the end. . . . I consider Oliver a little bit of a lucky charm. You need a little bit of luck to succeed." Mark Giles, a prominent banker in Charlottes-ville, also kicked in. The son of an engineer, Giles had found him-self captivated by the audacity of Oliver's plot to win an electric-car competition with a gasoline-powered vehicle: "I told Oliver, man, if you win this thing, you're going to be the Antichrist."

Today, though, in early April of 2010, Edison2 is out of money. The initial funds have been spent. The company is $63,000 in the red. Oliver needs the $1.3 million bank loan to pay his mechanics and employees and to keep buying supplies.

Now Oliver drives to Jarman's Sportcycles, the motorcycle shop outside of town. He tells me he needs to pick up some parts to take to the Edison2 workshop. Navigating through the rows of bikes, Oliver stops at a Yamaha with a blue-painted frame and gazes for a second at the bike he has cannibalized to build the Very Light Cars. "This is what we buy," he says. "It's beautiful." He smiles and cocks his head admiringly. "The way this hub is made," he says, running his finger along a smooth silver bulge, "the way this radiator is made, is the way a car could be made. But cars aren't made this way. They're just cruder, heavier, bigger." Pause. "Isn't that funny? It's a motorcycle, *in our car.*"

Oliver approaches the register to see if the parts are ready. He assumes someone from the shop has already called ahead and paid for them. An older woman behind the desk smacks her gums and mentions something about a past due account.

"Is this not paid for?" Oliver asks.

"No, honey."

Oliver fumbles with his wallet. From a back room, a gray-haired man emerges, then approaches the counter. He looks down at Oliver's wallet and up at Oliver's eyes.

"Should I close this out?" the woman asks the man.

The man nods to the woman, who hands Oliver his cardboard box of parts.

A few minutes later, heading southwest on Route 29, Oliver receives the first of several calls from creditors. His voice settles into a soft coo, a series of low tones and clicks, as if he were coaxing one of his children to sleep. "Paul, yes, I don't have my big check for you. Call me tomorrow at 8 A.M. if you think of it and I'll have George pick it up and run it over to you." (George works for Oliver; he makes coffee for the Edison2 team, does odd jobs, sweeps the shop floor.) "Larry, I'm going to have to hold on to what I owe you. . . . We might end up drumming up a fair amount of business for you."

Oliver stops for gas. He takes out his wallet, then shoots me a panicked look.

"Do you have any cash?" he says.

IN 2009, Oliver attended a conference in New York about the automobile of the future. He didn't know anyone in New York, so instead of carrying a briefcase from session to session, he carried one of the Very Light Car's custom wheels. At various points in the conference, people approached Oliver to ask what was up with the wheel, just as he had hoped. As Oliver chatted people up, he told one guy, "The great thing about being a small, nimble company is that I can do with 5 million what GM can't do with 50 million."

The guy said, "You know, you may be right." He gave Oliver his card and walked away.

Oliver looked at it. "Stephen Girsky," it said. "Vice Chairman, General Motors."

A flutter of e-mails followed, resulting in an invitation for Oliver and his team to visit the GM Technical Center, located twenty miles north of Detroit, and explain to some of the company's engineers why the Very Light Car represented the future of the automobile. The meeting would take place in May, immediately after the X Prize's Shakedown stage. Oliver would have thirty minutes to make his case to the second biggest car company in the world.

10

Edison2 Welcomes Informed Comment

LYNCHBURG, VIRGINIA, APRIL 2010

"We are out of time," Oliver says. "We are out of time in a serious way."

He frowns and veers off to take care of some urgent task, his boots clacking loudly across the workshop's pine floor. Shakedown begins in six days. The cars have to be finished by then, or else it's all over for Edison2. The rhythms of the shop have quickened since I was here last month—there's more tool noise, more chatter, more bodies in motion. Oliver has called in favors from several old friends and colleagues, persuading them to come to Lynchburg and help out in this desperate, last-minute push to get the cars ready for the track.

Amid the commotion, though, is a zone of stillness. At his desk, wearing a crisp blue shirt and khakis, Ron Mathis opens his laptop. He clicks several times, and the space-egg curves of the

Aptera 2e fill the screen. It's been six days since Aptera revealed its newly redesigned car, and Ron has been examining pictures for clues to how it might perform in the Prize. He nods and steeples his hands.

"It looks to me like it's a completely different car," Ron says. "We're impressed that they expect good aero with the shape they have. We think the wheel pods are shit. They're going to make the air angry." Ron points to a spot on the hindquarters of the 2e: "I would expect the air to burble right . . . *here*."

Ron thinks Aptera has fallen into a trap: By trying to make their car more appealing to future consumers, they've also made it heavier and clunkier, harming its chances to win the competition. He admits, though, that there are some nice touches on the 2e. "This car, it's clearly going to present better than we are," Ron says. "Look at the interior. It looks nicer than ours. Look, they've got cupholders." A naughty grin flashes across his face. "I didn't think of that. We might make a fancy machined one just to piss them off."

"What I see here," he concludes, leaning back in his chair, "is something that looks a bit—ordinary."

As for the Very Light Car, Ron says, "We're looking pretty good in lots of ways. Touch wood so far. We've had very few mechanical issues with the car." He beckons me over to one of the prototypes and says he wants to show me something.

The men of Edison2 understand that the car is a disorienting object, which is why they never try to explain it all at once. Instead, they begin by pointing out a single small part of the car that exemplifies the whole. For Oliver, that part is the very light lug nut—the tiniest, humblest part of the car. The irreducible kernel of a philosophy. For Ron, the crucial part is something a little larger, a little more complex, a little more obviously beautiful.

"Have you seen our hand brake?" he asks. "There are some flourishes."

He encourages me to sit in the car so I can more fully experience the hand brake. I begin to climb in, placing my right foot on the cabin floor. The car rolls forward several inches, just from the pressure of my foot—it's that light. I have to crawl in delicately so as not to push it several yards across the floor.

The interior is black as the char on a steak, with exposed steel tubing—an unfriendly, high-performance space. It feels like sitting inside a piece of stereo equipment. The hand brake looks like something out of a high-end erector set. Most hand brakes are ugly afterthoughts; anyone would instantly perceive this hand brake as something different, an object carefully wrought. It was designed with Ron's CAD software and created by Brown Machine, the Lynchburg shop. The brake lever, made of aluminum, has six little circular holes in it, to shave weight. Its color is pale gold, the result of a chemical treatment, called Alodine finishing, that made its exterior resistant to corrosion. But that's not all. Inside the brake, Ron explains, instead of the usual steel cable that runs to the brake piston, there's a pushrod connected to a small reservoir of hydraulic fluid. The fluid in the accumulator compresses and expands according to the action of the brake lever, so when you lift the lever, the motion is completely smooth.

Why go to all that trouble just to lighten and smooth out the motion of a brake lever? "It's nice to take pride in what you do," Ron says. "It would be a sad world if you couldn't do something nice."

Once I understand the hand brake on a deeper level—the thought that went into it, the almost lunatic overflow of attention and skill—every inch of the Very Light Car starts to take on a new significance. The shifter knob, for instance—right there next to the brake lever. Magenta in color, Swiss-cheesed with holes, it weighs two-tenths of an ounce. And every part of the car is like this, even the parts that are tucked into deep and inaccessible crannies—parts that no driver and very few mechanics will ever even see, designed by an engineering mind so uncompromising that if you were a competing X Prize team and you poked your head in the door you might start to despair.

"I hope some people would look at it and give up in their heads," Ron says.

OLIVER'S STRATEGY of overwhelming force—his decision to build four cars instead of just one or two—has saddled the small team with a nearly impossible workload. On the one hand, the four Very Light Cars are essentially duplicates of one another, except for the number of seats, so it's not as though Edison2 is building four unique cars. On the other hand, each car is a prototype, and each prototype has its quirks; as Ron puts it, "they're basically snowflakes."

Down the hall, in the body shop, it becomes apparent just how far behind schedule Edison2 really is. Hard rock plays on the radio as mechanics Bobby Mouzayck and Reg Schmeiss sand away at composite body panels, scrambling to complete two cars—one of the two sedan-like mainstream cars, plus the two-seat tandem. The mainstream car has a body but no windows and no wheel pods. The tandem doesn't even have a body yet. "It's nowhere near complete," Reg says. His T-shirt says GUT WRENCHIN' SHORT CHUTES, WHITE KNUCKLE STRETCHES, 210 MPH RACIN' AT THE YARD. He absently digs into his finger with the tip of a razor blade without looking down at his hand. "It's a splinter," Reg says. "I get thousands of 'em every day. Dunnworryboutit."

Bobby and Reg seem unimpressed by pretty much everything around them. They joke about how much it sucks to live in Lynchburg. The defining institution here is Liberty University, the biggest evangelical Christian university in the world, founded by the preacher Jerry Falwell. "In the rest of the world," Reg says, "they're having fun and crazy unprotected sex," but not here. "Lynch

Vegas," the mechanics call it. They gripe to me about how the engineers make their jobs difficult. "You could fill twelve chapters in your book," Bobby says. I tell Bobby and Reg that they seem like professional pessimists, and Bobby nods. "You kind of have to have an 'it's all shit and they're all cunts' attitude just to keep it going. In every rivet you put in, you can space them properly and make it nice, and we expect the same from the engine people and from Oliver on down to George.

"Races are won at the track and lost in the workshop," Bobby goes on, "and I don't think many people understand that. Who knows, that Thailand piece of shit that's made of papier-mâché, with the similar engine program as we have, they may fuckin' win." He's talking about the Thai team that used a pig carcass to crash-test their foam car. "The car is a doghouse with four wheels. You know what I mean? It looks like the first bobsled ever made. They may pull it off, who knows? They have no shortage of labor. We don't have enough good guys."

Back out on the shop floor, a man arrives from the local machine shop where most of the car's parts are made. Kenny Brown has brown-and-white hair and a thick mustache. His green polo shirt says BROWN MACHINE.

The existence of Brown Machine is one of the reasons Oliver decided to build his cars in Lynchburg and not some other town. The same quirk of geography that makes it lonely to live and work here also makes it a great place to build the prototype of a new machine. The inventors of the cigarette-rolling machine, the at-home enema, and ChapStick are from Lynchburg. In the twentieth century, residents seeded a vibrant textile industry along the James River; the textile factories depended on a series of local machine shops to keep them supplied with replacement parts, since it would take too long to order parts from other cities. And when the textile factories closed, the machine shops were saved from certain doom by the presence of one of the biggest companies in town, Babcock & Wilcox, a leading manufacturer of nuclear-power plants. Babcock also supplies spare parts for the U.S. Navy's nuclear submarines—and Brown Machine makes many of those parts, on the same computerized milling machines it's been using to make parts for the Very Light Car.

Oliver introduces me to Kenny Brown. "Listen," Oliver says, "I'm very open. One of the reasons he's here is that I owe him money. But he has never once pushed me." The reason Brown hasn't pushed is because he sees the bigger picture, Oliver says. He starts talking about how devastating it will be for the wider economy if a place like Brown Machine goes under. You would lose not only the shop, with all its milling machines—you'd also lose the expertise required to operate them, dispersing a generation of trained machinists. If you do this in enough cities, you'll hobble your ability as a nation to innovate.

"An industry can't be born without the backup materials," Oliver says, "and the backup materials can't be born without a shop like ours. The idea that you can draw it here and end up making it in China? Just ain't going to happen. You have to draw it here, try it, play with it, modify it, change it." He reaches into his pocket and holds up a lug nut. This is probably the third iteration of the lug nut design, Oliver says. "It ain't lug nut number one. By the time you mass-produce it, you might be producing number seven."

Kenny Brown nods through Oliver's monologue. He tells me he's been stiffed by clients before, but sometimes you have to go on faith. For the money Oliver owes Brown, either one of them could buy a decent house. Brown speaks in a drawl so deep it dissipates almost completely in the squirts and bleats of the shop's hydraulic tools. "If I get snookered, I get snookered," Brown says.

Oliver looks at the floor and mutters, "That's not going to happen here."

———

THAT NIGHT, Bobby stays up late, working on the windows of the four-seat car. "I guess I could schlock together some crap," he says, "but then it's crap."

Eventually, around 3 A.M., eyes watering from the carbon dust, he decides to get a couple hours of sleep. He walks toward the back of the body shop, passing a rack full of carbon scraps that, thrust together randomly in piles, look like scale models of Frank Gehry buildings. He continues into a vast, unlit storage room littered with Oliver's junk, the residue of old passions: a stack of art deco posters; a claw-foot bathtub full of skis, sticking out like the limbs of drowned corpses; a gaudy, starving-artist painting of a man in a forest. Then Bobby approaches a door at the side of the room. It's unlocked. He opens it, entering his apartment.

The room is small and sparsely furnished. There's a pull-up bar near the bed, a tiny kitchen, and a surprisingly nice bathroom with Moroccan tile. On top of a bureau is a picture of a young Bobby with his father, Robert. A pack of mostly empty Marlboros rests against the picture—the last pack Robert ever smoked.

Bobby got the news just last month. His father, despondent over his gambling debts and the failure of his automotive business, had killed himself. "I thought for sure I was over the hurdle of character-building," Bobby says. "I was like, oh, it's just gonna be easier now. And it's amazing what you take for granted." The morning he heard about his father's suicide, Bobby told Ron he needed a few days off from Edison2 to drive down to Atlanta for the funeral. Ron told him, "I'll fly you. Listen, you're not fucking driving right now." Ron flew Bobby to his father's funeral in his own personal plane. "That was the coolest thing, period," Bobby says. "No matter how mad I get at Ron, I'll never forget that."

Now Bobby falls asleep in his clothes.

Four hours later, at 7 A.M., Bobby wakes up and limps out his front door into the shop without showering or shaving or changing. His curly black hair is flat and matted against his scalp. He starts back on the body panels, sanding and sanding, until, after a few hours, he can no longer see. He throws his head back and squints violently at the ceiling, trying to weep the carbon filaments out of his eyes.

At this moment, P.K. happens to walk past the body shop. Seeing that Bobby is in distress, he grabs a nearby shot glass and some contact-lens solution. He cleans the glass in a sink, fills it with lens fluid, and hands it to Bobby. Bobby pries open his right eye with his thumb and forefinger, tilts his neck back, and splashes the shot into the exposed white pulp. He does the same with the left eye. Tears and saline stream down his stubble. Then Bobby walks out of the body shop and into the main room, pours himself a cup of black coffee from the pot near Ron's desk, and adds enough sugar to make a cake.

"WE THINK they're fucking idiots," Ron tells me. He's at his desk, paging through a document that just arrived by e-mail from X Prize headquarters. "Every time they put pen to paper, they prove it."

The document is thirty pages long. It details, for the first time, many of the safety features the cars will need to demonstrate in a "technical inspection" before they'll be allowed to compete in Michigan five days from now. "If they've got an army of little Hitlers there, looking to find fault, it's going to be difficult," Ron says. "But there are plenty of internal contradictions in the stuff that they've sent us that we've had to deal with."

Ron has been doing battle with the Prize officials for more than a year, trying to make sure they understand—and don't penalize—the more unusual features of the Very Light Car. His chosen weapons have been the series of technical submissions Edison2 and all the other teams have to ship off to the Prize every couple of months. The submissions are meant to be plain, matter-of-fact. But Ron's submissions are things of beauty. They argue, threaten, and charm. They invoke historical anecdote. They lay out stark positions and defend them with a coiled rage masked by extreme formality:

[X PRIZE] REQUIREMENT: A discussion of how the vehicle or vehicle modification is designed to achieve the normal bumper-to-bumper warranty of 3 years or 36,000 miles on all major components and a 10-year, 120,000 mile emissions system warranty.

[EDISON2] RESPONSE: What is a warranty? It is not a guarantee a vehicle will run for 3 years/36,000 miles without trouble of any kind. It is a guarantee the manufacturer will fix the car if it does not. In the 1960s, Lotus founder Colin Chapman explained to his bankers that his company's fantastically high warranty claims were really a saving because they were the result of money Lotus had not spent on development.

[X PRIZE] REQUIREMENT: Fuel System Safety. With reference to the schematic required in tab 1, identify potential safety hazards of such systems and describe in detail how the fuel system addresses each potential hazard.

[EDISON2] RESPONSE: While it is obvious the price of motion is the risk of collision, Edison2 does not concede its cars contain hazardous design features, either generally or in the fuel system in particular. . . . Edison2 personnel have between them numerous first class international race and championship wins. We have built fuel systems before and done them properly and we will do so again on our X-Cars. It is not possible to distill the decades of combined knowledge and experience that have been applied to our car into a brief technical paper. As we have previously said, informed comment

and discussion are welcome and [X Prize] personnel
are invited to our shop to see for themselves.

Many of the exchanges boil down to Prize officials asking Edison2 to prove that the Very Light Car is safe in the manner of a "normal" car, and Ron then trying to explode the assumptions inherent in that request.

Is the car safe?

Well, it's got DOT-approved tires, a shatterproof windshield, a wide stance, a low center of gravity, and strong brakes. Its tube frame is made of steel, and the frame is bolstered by a body that contains semi-structural composite materials.

Is the car safe?

Well, come on, let's be honest, it's a *prototype*, designed to win a competition and prove that a certain ridiculously high level of efficiency is feasible. Additional safety features can be added later, once the main engineering issues are worked out. This is about building a platform, not a finished car.

Is the car safe?

"We refuse to speak in absolutes like this."

Is the car safe?

Define car. Define safe.

WHILE RON battles the X Prize officials and the mechanics battle Ron, Oliver tries to be helpful to everyone. He strides around the shop with a twenty-five-foot tape measure clipped to his jeans. He periodically sucks in huge, satisfied breaths as if he were inhaling fresh mountain air instead of a fatal fog of aluminum, carbon, and epoxy dust.

On the original Very Light Car prototype, the silver car, Oliver personally performed a lot of the welding. It's unclear whether Oliver knows that these silver-car welds have since become a subject of much amusement in the shop. As fond as the men are of Oliver, they don't respect his mechanical ability. "Absolute crap," says Bobby, "no two ways about it." The welds on the silver car look so awkward and ugly, the joke is that Oliver probably went out to the trash heap in back of the factory and grabbed an old sewer pipe and welded it to the braces.

Around 5 P.M., Oliver walks back into the body shop. For some reason, Bobby isn't there. Maybe he's taking a smoke break, or maybe he's in some other part of the shop. Whatever the case, Bobby doesn't see what Oliver does next until it's almost too late. Oliver lifts the body panels of the tandem car. He carries them out to the main room and places them on the tandem car's chassis. He's trying to speed up the job. But Bobby has been working three weeks to prepare the panels, and he has a specific, correct way he wants to attach them to the chassis.

When Bobby returns to the body shop, he sees that the body panels are gone. Then he sees what Oliver did. Then he starts to yell.

"Dude, do not fuck us like that," Bobby tells his boss. "You'll have a car that rolls but DO NOT fuck us like that."

People stop what they're doing and stare.

Oliver puts his hands up. His face clouds, slackens. He responds to Bobby with a sober, almost courtly formality. "I did not know that," he says, his mouth forming an O of retreat. "You did not communicate that to me." He turns his back and walks away.

Bobby storms into the body shop, muttering under his breath. The words "duct tape and fantasies" are audible. He asks P.K. if he has any money. P.K. hands him a crumpled bill and Bobby walks away. "Money for his milk and Snickers," P.K. says. He smiles, then adds quietly, "Well, that cleared the air. Put the boss in his place. That's how we do things around here."

Work continues into the evening. A hundred and ten hours until Shakedown. The sun sets, but there's no way to know; the

light in the body shop remains a steady fluorescent yellow. Bobby
and Reg are crawling atop the tandem car. In the last three hours,
they've fitted the body panels onto the car. Now they're sanding
away excess glue along the seam where the panels come together,
a seam that runs front to back along the top and is masked off
with lime-green tape. It's approaching 8 P.M., but instead of slowing down at the end of a long day, the mechanics are speeding up.
"Yeah," Reg says, rubbing vigorously with a piece of sandpaper,
"we're kicking ass. This is how you do it."

A few minutes later, several large pizzas are delivered. The
engineers and mechanics of Edison2 congregate on the main shop
floor, clustering around the two four-seater Very Light Cars. They
grab pizza slices and pile them on paper plates. The machines fall
silent, one by one. Carbon dust settles on the cheese.

The racers stare at the cars and one another.

"It's so funny to see the cars like this," David Brown, the communications chief, says, finally, taking a bite of his veggie slice.

"Funny?" asks Brad Jaeger, the young engineer who sparred
with Oliver back in March regarding the car's aerodynamic drag.

"I don't know," David says, chewing thoughtfully. "It's almost
like seeing a new beetle species descend, or something."

"Well," Ron says, "it's a flotilla." He pauses. "Soon to be an
armada." A giddy, maniacal look flashes across his face. "It's the
biggest make-work project in the history of projects."

Make-work. Ron is right. No one forced the men of Edison2
to enter a contest that could so easily expose them, instantly, as
bullshitters and fools. Everyone in this room has staked his professional reputation on what will happen at Shakedown. Several have
staked much more. And if Edison2 fails, it's unlikely that anyone
will be too forgiving: not the investors, not the Prize officials, not
the media. Or even, for that matter, Nature itself.

The Team Favored By Physics . . .

11

The Moose Test

To get to Shakedown from where I live, in Philly, I've got to drive ten hours northwest, across all of Pennsylvania and Ohio and into rural lower Michigan, sneaking beneath Detroit and entering a lake-spotted region called the Irish Hills: cabins, campgrounds, barbecue shacks, convenience stores advertising live bait. Past a mini-golf course cut into a hillside, I come to a large open field and glimpse the Speedway for the first time, a massive Cheerio dropped into what looks like a cow pasture, with no other structures around for hundreds of yards. Twice a year, for NASCAR races, this area is packed with camper vans, recycling containers overflowing with empty beer cans, and 100,000 people from all over the country. Today it's a ghost town.

I show my ID to a security guard at the Speedway's gate and drive toward the building on the outskirts of the track where the teams have been told to register as soon as they arrive. Inside, I spot a guy with an X pin on his lapel: Eric Cahill, the Prize's executive director. A sober, analytical man of thirty-nine—he used to be an intelligence officer in the Navy, then studied at MIT—Cahill says he's just getting his first look at some of the cars, and that what he's seen has both inspired and troubled him.

"If you're gonna engineer and build stuff, you need to know about markets and distribution partners and channels to get the product, once it's designed, out there," he says. "And that's one of the things I worry about with these teams. Not all of them, but some of them, certainly. That they've gone to great lengths to make a competitive vehicle—technically, a credible and competitive vehicle—but will it be successful in the market? Because plenty of great ideas never get there. And interestingly, plenty of subpar ideas *do* get there."

There are only twenty-eight teams left in the competition. Cahill says he's happy about how the teams have performed so far—he expects five in each category to make it all the way to the finals—but he allows that there could be mishaps. Recently, Cahill and his colleagues got word that a British team, Delta Motorsports, had suffered a catastrophe. They'd plugged their electric sports car into a charger and left it in a garage overnight. The charger didn't shut off when it should have. "It just kept charging the batteries until they melted," Cahill explains. "Caught fire. It's the old thermal runaway. Took down their car, a good chunk of the garage. We've got some great photos. The car looks like a burnt potato chip. It literally looks like it spontaneously combusted. It's just like . . . ash."

THE NEXT morning, I pick up the local paper. It doesn't mention last Tuesday's catastrophic explosion on the Deepwater Horizon, an oil rig in the Gulf of Mexico. The explosion sank the rig, killing eleven, and on the ocean floor, a mile beneath the surface, the broken well is now pouring 42,000 gallons of oil a day into the Gulf, with no sign of stopping. There's only local news: an article about a seventeen-month-old boy on the verge of death after ingesting methadone, one about scientists at a local university asking hunters to donate wild turkey hearts for research purposes, and one about a rally at Vandercook Lake in support of open-carry gun laws. But nothing about the Prize. I enter the Speedway through a different gate today, dipping down through a tunnel and crossing the track to emerge on the infield, where several bulldozers are shoveling mud into big piles, part of a renovation project the Speedway staff has inexplicably scheduled to take place during the X Prize. If it wasn't already obvious that the Speedway makes all of its money from NASCAR races—that it's only indulging the X Prize Foundation to get a bit of good press—it's obvious now.

On one side of the infield are three long, multi-bay garages where the teams have unloaded their cars and tools and have begun to eye one another. I see Edison2's Bobby Mouzayck standing on the lip of one of the garages, smoking a Parliament Light.

"Well, we did it," he says. "Things are pretty precarious, but fuck it." He points to one of the four Very Light Cars. "This one's a bag of shit." He points to another. "*This* one's a bag of shit."

Bags of shit and bulldozers notwithstanding, Bobby is glad to be here, he says. After a year of talk and CAD drawings, a year of planning and machining and casting and cutting and sanding and scraping, he's finally going to *race*, albeit in an unusual and possibly diminished way. He turns to face a nearby vehicle, a twelve-foot trailer with a picture of two cartoon pigs on the side. NORTH CAROLINA NATIONAL CHAMPION COOKING TEAM, it says.

Bobby says he knows Oliver doesn't have much money left for niceties. The pig trailer is the best Oliver could do. But in his opinion, if you're going racing, you don't show up in a used trailer from a barbecue joint. In real racing, you show up heavy: twelve guys in identical Kevlar jackets spilling out the back of a tractor trailer. Bobby would have preferred something more like the Aptera Motors trailer beyond the far end of the garage. It's a proper fifty-three-footer with a big-ass logo on the side. People in identical white-and-green uniforms duck in and out of the trailer with purpose, smelling of money and fear. Aptera has come heavy.

That's racing.

Bobby takes a puff of his Parliament. A gust of wind blows the smoke almost straight up. He looks to his right and sees Illuminati Motor Works: four people in lawn chairs, wearing jeans and baseball caps and untucked polo shirts, drinking diet sodas—as if this weren't a race for $10 million but a fucking backyard cookout. And behind them, their car, which he calls "that limousine." That's the first thing that springs to mind: "A hundred-mile-per-gallon limousine," he says. A long and curvy monstrosity. Huge, garish fenders, gullwing doors that don't quite fit into the holes that have been cut for them. Its color is the raw light gray of primer coat. The hood is up, spilling complicated guts: wires, zip ties, plastic buckets, and what looks like . . . no, *couldn't* be . . . Is that *plywood*?

Jesus.

Maybe, if you've built a car like that, you should start off at the county fair before you come to a real race?

"You can't show up like pussies," Bobby says, stubbing out his cigarette. "You have to show up with an iron fist."

He looks again at the limousine.

"Oh my God. Doc from *Back to the Future* did better."

AS I walk through the garages, past a disorienting array of tool cases, welders, cables, laptops, overnight bags, stacks of technical manuals, coffee cups, and vehicles in dozens of shapes and colors, I notice people stealing glances at the Very Light Car. Until now, they've seen it only in pictures.

"They are basically a four-wheeled motorcycle," says Michael "Doc" Seal, the founder of a team from the vehicle research program at Western Washington University, whose student engineers have traveled here in a Winnebago. The WWU car is a hybrid coupe called Viking 45, after the school's mascot. Seal looks the Very Light Car up and down. "I think for many people that will be a really nice car." By "people," he means motorcyclists. "Motorcyclists are not concerned about crashing into a concrete wall at any speed. They know they die." Seal adds that he's proud that WWU crash-tested a close predecessor of its X Prize car "with mannequins and everything," adding, "The X Prize people aren't just looking for cars that get exceptional economy. That, by the way, is not very difficult. It *is* difficult to do it in a real car that Americans will buy."

In general, the more conservative and market-oriented a team's vehicle, the more ridiculous that team finds the Very Light Car. One of the simplest cars in the competition belongs to BITW Technologies, a team from Indiana that has converted a Chevy Metro to use a biodiesel engine usually found in lawn-care and small industrial equipment. "Some of the cars are very . . . imaginative," one of BITW's leaders tells me. "Like you think you're in Dr. Seuss." He leans in closer. "Just read *Green Eggs and Ham*. You'll understand. You'll think you're Sam-I-Am."

OLIVER FEARS only one of the teams he sees in the garages: X-Tracer, from Switzerland. X-Tracer has entered two banana-yellow vehicles it calls E-Tracers. Powered by electric motors and battery packs, they seat two people, front to back, and resemble motorcycles, with a few key differences. One, they've got lids—shells of Kevlar that enclose the driver. Two, whenever an E-Tracer slows to a stop, a set of training wheels flicks out to each side, balancing the vehicle. This helps the E-Tracers meet the Prize's "drivability requirement." According to the rules, all vehicles "must be able to be driven by an independent driver with average driving skills given only ten minutes of instruction/training," and presumably a driver with average skills wouldn't know how to drive a motorcycle without tipping it over. Hence the training wheels.

Still, the E-Tracers are basically motorcycles, and some people here have been grumbling about their presence in the Prize, including Oliver. An electric motorcycle has obvious advantages in an efficiency competition: It has two wheels instead of three or four, so it's skinnier and lighter and displaces less air. To Oliver, the E-Tracers, which are competing against his tandem version of the Very Light Car for the $2.5 million prize in that division, seem like a cheat—less a sincere solution to a problem than an attempt to slip through a loophole in the rule book and grab the cash. How many Americans are really going to give up their SUVs and Priuses to start driving a motorcycle? (Later, Oliver tells me he ended up doing the gallant thing, arguing to X Prize officials that the E-Tracers actually did conform to the rules and should be allowed to compete.)

Oliver gets along, though, with the creator of the E-Tracer, Roger Riedener, which isn't surprising. The two men are almost doppelgängers. A jowly man with black-and-white stubble, Roger makes his living in real estate. His passion is building and racing radio-controlled airplanes. Like Oliver, he is ardent about lightness.

"Have I told you the tale of the hummingbird and the swan?" Roger asks me, in a confident, soft yet throaty voice. "The hummingbird is a quarter of an ounce. A full-grown male. And a quarter of that is its wing muscles. And with those two grams of wing muscles, it is able to sustain hovering flight over the stem, when it drinks the nectar. Now, you take a swan, and you tell it to do the same party trick—the swan is twenty pounds, and it's a fact that it would need eighty pounds of wing muscles to do the same party trick." He smiles slightly. "We are the hummingbird." Roger says that several of the electric cars in the X Prize—especially the Aptera 2e—are "over the edge. They are already small swans. Young swans are called cygnets, I think, in English?"

He pulls out a pack of Camels, removes a cigarette, and holds it between his fingers. He says he makes no apologies for the E-Tracer. He doesn't understand the complaints. The rules clearly allow such a vehicle: "It's just the smallest damn nutshell that you can fit two people in, and the largest motor." As for practicality, well, his car and Oliver's only appear like extreme departures from other cars on the road "if you look at them with the eyes of the horse-carriage drivers." In the future, Roger says, American drivers will have to accept fewer creature comforts as the price of driving more efficient cars. "Seven cupholders? Air-conditioning that even in 105 degrees cools down to 65? Mmmmmmmm." He lights the Camel and takes a puff.

IN REAL racing, you drive the course before the race. You examine every crack in the asphalt, every dip and deviation from smoothness. You map out how your driver will take advantage of the little imperfections in the track to make the car go faster. Oliver jangles the keys to a Ford Super Duty van and gestures to Ron and Brad to climb aboard. He's setting out on a reconnaissance mission. He wants to check out the fields of battle. Surely the Prize officials won't object if he takes a leisurely drive around the facility?

The three men drive the van south across the infield, past the low cinder-block building where the Prize officials have congregated. They pass through a chain-link gate and cross a parking lot where the Prize has set up its official fueling stations beneath a long white tent. Past the fueling tent, Oliver takes a hard left and circles back to the open, grassy part of the infield where no competitors have yet ventured. On a small S-shaped access road, officials have set up a series of speed-limit and stop signs. SPEED LIMIT 50. STOP. SPEED LIMIT 15. SPEED LIMIT 45. This is the optional-for-now "emissions" course. Later in the week, each car will have a chance to drive the course while towing a trailer containing instruments that sniff the chemicals coming out of the tailpipe.

Oliver knows that not all of the tests at Shakedown—and at Knockout and the Finals to follow—will be held on the two-mile NASCAR oval itself. Some will take place on courses both inside and outside the oval.

A dark silver Prius now approaches Oliver's van. The Prius contains four Prize officials in white polo shirts. They don't look happy. Oliver rolls down the van's window and pokes his head out. He's about to get taken to the principal's office.

"Steve told us we could drive around the course," Oliver says, invoking the name of Steve Wesoloski, the Prize's technical director.

"Did he use the words 'around the track'?" asks the Prize official.

"He used the words 'around the track,'" Oliver says, giving no ground.

"Did you see Aptera driving around the track?"

Oliver mutters a vague apology and says he's going to leave now.

But he doesn't return to the garages. Instead, he steers across the infield to the far side of the oval, and then to a road that leads outside the track, toward an older, shuttered part of the Speedway—a curvy test track that has cracked and split in the Midwestern sun. In other words, it's a perfect place for the X Prize to test the durability of the cars. According to the Prize rules, for a prototype to advance in the competition, it has to complete twenty laps here without breaking down.

On the durability course, Oliver bombs up and down hills. The van sounds agonized. He takes the turns sharply, spraying gravel.

Having gotten a good look at the terrain, Oliver now heads over to the far straightaway of the oval. All three men get out. They stare down the straightaway, about half a mile, at two rows of orange traffic cones spaced eight feet apart. The lane created by the cones runs straight for a time, then makes an abrupt S-shape, then returns to straightness.

The Prize is calling this test an "avoidance maneuver," or a "double lane change," but it's known informally in the auto industry as the Moose Test. Imagine you're driving along at speed and a moose suddenly jumps out into the road. You've got to swerve once to avoid the moose, then again to get back into your lane and avoid oncoming traffic. Over the years, some of the most shocking revelations about auto safety from *Consumer Reports* have emerged from the Moose Test. If you've ever heard a news story about an SUV flipping over in a curve, chances are it happened in a Moose Test. And this is the industry-standard course. The Prize has actually hired *Consumer Reports* employees to administer all of the dynamic safety tests outlined in the contest rules, including tests of acceleration and braking.

To pass the Moose Test, each car will have to negotiate the cones at 45 miles per hour without using the brakes or the throttle.

Oliver turns to Ron and shakes his head.

"It's pretty narrow, man. That's 8-foot-wide at 45 miles an hour." He looks back at the van, then at the cones again. He raises his eyebrows. "You wanna try it in the van?"

Brad coughs out a clenched little laugh. "*I* wouldn't do it," he says. And he's the professional driver.

"I know you wouldn't," Oliver says. "That's why *I'm* gonna do it. You want to ride with me?"

"I'd rather . . . not."

Ron pipes up, "*I'm* going."

Brad shrugs and runs after Oliver so he won't miss the action. Oliver pulls on his seat belt and turns around to face Brad in the backseat.

"Not like I haven't done it before on the highway," Oliver says.

"I don't know about that," Brad says.

"This is going to be funny," Oliver says.

With a quick nod to Ron, Oliver stomps on the gas pedal. The van accelerates into the cones. First at 30 miles per hour, then 40. When it get up to 40, it starts to jiggle. At the first swerve in the cones, Oliver pulls hard on the wheel, then hard again. The van shudders. It feels ready to tip. He coasts to a stop.

"I'm not going to roll a van today," Oliver says, slowing down.

But then he doubles back to the start and goes through the course again.

The speedometer reaches 45 as it enters the S. The van rattles as it pulls left, then right. Oliver barely makes it to the far straight-away without knocking over any cones.

He sucks a great deal of air through his teeth.

"This is *not* trivial. Most of these cars aren't going to be able to do it." He's thinking particularly about the Aptera 2e, which has only three wheels and ought to be less stable in the curves. "Honestly?" Oliver says. His grin thunks on like a bank of stadium lights. "This could be a *huge* problem for Aptera."

MEANWHILE, BACK at the garages, the men of Edison2 are preparing the Very Light Cars for the track. All told, there are thirteen team members—Oliver's full-time crew from the Lynchburg shop, plus several additional engine and transmission experts Oliver has hired to make sure everything goes smoothly. They wear identical slate-gray shirts detailed with the alien-green Edison2 logo and drink coffee to stay awake. I check in with Bobby. He says he's unhappy with the work ethic he's seeing in some of the guys. Last night, he says, when they all stayed late in the garage to work on the cars, "they're like, 'We're hungry.' Why the fuck do you think you're here? It's not to eat. It's the winning thing. We take our toy out of the box. We show that our toy's better than their toys.

"And another thing. Stop drinking coffee. Start with the amphetamines already."

IN A garage bay reserved for technical inspections, Simon Hauger of the West Philadelphia Hybrid X Team stands next to one of the team's two cars, answering an endless stream of questions from an elderly man in a tan polo shirt with an X above the left breast. The students aren't here at the moment; if West Philly survives Shakedown, Simon plans to bring them to Michigan for Knockout.

The man in the tan polo is retired from Chrysler. He holds a clipboard containing a twenty-nine-page inspection checklist. It covers areas such as cargo capacity, chassis and crash safety, seat belts, egress, bulkheads, fluid leaks, ground clearance (the bottom of each car has to be at least 4 inches above the ground), and vehicle noise ("all vehicles will be required to meet drive-by ambient noise standards of 74 dB max"). There are pages and pages of electrical requirements. Any electrical conductors "that may be damaged by moving parts, bending, chaffing [sic] on corners or surfaces, pinching, crushing, high temperatures, or corrosive liquids" have to be protected by layers of insulation. Each electric and hybrid car is required to have an emergency disconnect switch—essentially a "panic button" that will shut off the car in case of a fire or other catastrophe—and there are ten separate checklist items covering the emergency disconnect switch alone.

The man in the tan polo shirt asks Simon what torque spec he used to tighten the bolts on West Philly's car.

"The German one," Simon says.

"What is that?" asks the man.

"All the bolts are *gutten*-tight," says another West Philly adult, "Big" Mark Dougherty.

"Funny," the man replies, not smiling.

Hauger and Big Mark are in the inspection bay for another five hours.

Team Illuminati at Shakedown. From left: volunteer Patrick O'Ravis, Nate, Kevin, Nick, and George. The man on the far right is Oliver Kuttner.

THE NEXT morning, the Speedway is a bowl of cold, still air. The teams mill around the garages, waiting for the tests to begin. Aptera, Edison2, and West Philly have passed inspection. Kevin Smith's team has not.

He's here with his wife, Jen; his father, Nick; and Nate, the electrician. The judges have just demanded that he and his colleagues add several layers of insulation to their electrical system. Nate and Kevin lean into the exposed dash, pulling at tangles of cables, struggling to make the fixes as quickly as possible.

An elderly Italian man named Joseph Randazzo watches Nate working. He's here as a volunteer with Edison2, along with his son, Tony. Joseph is a friend of Oliver's going way back, an auto-parts man from New Jersey. Decades ago, he fabricated replacement parts for some of the very first Ferraris sold in America. He knew Enzo Ferrari personally.

"I like the roofline," Joseph says of the Illuminati car. He says he's pretty sure the vehicle is a modification of some sort of European luxury car that isn't exported to America.

I tell Joseph that the car was made entirely by hand, in a cornfield in Illinois.

He shakes his head. "It's some kind of luxury car, definitely some kind of luxury car."

Having finished their work for the moment, Kevin and Nate button up. They close the hood, affix the dash, shut the doors, and place the steering wheel atop the steering column. Then they push their car back over to the inspection bay. Twenty minutes later, they push it back. Kevin makes a face like he's giving a presidential address. "One small step for man," he says. "One sleepless month for us." An X Prize official applies a green-and-white rectangular sticker to the windshield, signifying that the car has passed inspection.

Another official now comes up to Kevin with a concerned expression on her face. All along the judges have been slightly wary of Kevin's car, due to its handmade nature. They're worried that the tests will break it, and that Kevin will get hurt. The official says that several of the smaller, nimbler vehicles have had trouble with the Moose Test. She fears it will be a particularly difficult trial for Kevin, who will be driving one of the longest and heaviest vehicles at the competition. It weighs 3,200 pounds. She wishes Kevin luck and tells him to be safe.

The Prize required each team's driver to take a certified race car–driving course. The nearest one Jen could find was in California, and Kevin and his friend from work, George Kennedy, had flown out there, spending a day practicing controlled skids. Kevin completed the course. Still, he's nervous. "I've never done a lane-change maneuver before," he tells me. "Have you?"

THE WHITE car with three wheels is three-quarters of a mile away from the crowd of journalists and opposing team members that has gathered at the south end of the straightaway to watch the Moose Test. The Aptera 2e's front wheels are mounted inside aerodynamic pods. From this distance, it's hard to see the axles that connect the wheels to the body, so the car appears to float above the track, jiggling in the heat that rises from the asphalt.

A man in a white tent to the side of the Moose Test course waves a flag, and the Aptera 2e begins to grow larger. It enters the S. An orange cone topples, then another. The car seems to tilt slightly. The driver's-side door flies open.

OLIVER TELLS Brad to take the first run of the Moose Test at fifty miles per hour—five miles per hour faster than necessary. Oliver wants to make a statement. Brad doesn't want to make a statement, but he listens to his boss and does as he's told. He roars into the cones. He turns left into the bottom of the S, and when he swings the wheel back in the other direction, the tires begin to skid. Brad is spinning out. Instinctively he lets his foot off the gas and turns into the skid, straightening out the car. The car performs a perfect 180 and comes to rest facing the course it just failed.

Brad drives back to the start line and has another try. Within three attempts, he manages to navigate the course at the minimum allowable speed, 45 miles per hour.

Oliver, who's watching from the side, draws the conclusion from Brad's spinout that the minimum speed is also the maximum. He says, "You have to go at 45. If you don't go at 45, you don't get it."

Oliver and Brad return to the garages on the other side of the Speedway. For the rest of the afternoon, Oliver keeps one eye on the Moose Test, to make sure no team is showing him up. For a while, it seems as though he has nothing to worry about. Several teams, including West Philly, pass the test at exactly 45 miles per hour, but no faster. Also, Aptera is continuing to have trouble, just as Oliver foresaw; again and again, the white car spins out in the cones, failing to trace the shape of the S at speed. All in all, Aptera needs forty attempts to pass the Moose Test. (Because this is an early round, the judges are more lenient, giving Aptera an unlimited number of tries.) All of Oliver's predictions are coming true.

Then, at around three-thirty, he hears a rumor.

KEVIN IS going fast. Too fast. The S looms.

The *Consumer Reports* tester climbs atop a chair to take a picture of the car as it slams through the eight-foot opening. The tester takes a speed reading: 48.32 miles per hour.

The car maintains its speed through the S. It shimmies violently as it negotiates the turn. Then it cruises through the exit lane, past the last cone.

Kevin stomps on the brake pedal with both feet. Part of the pedal is a decorative door hinge. This is a recent improvisation. Earlier in the day, Kevin welded the hinge to the existing pedal to give himself more surface area with which to stop the car.

Kevin pops the car's gullwing door and climbs out. He removes his crash helmet and looks back at the cones.

He'd gone faster than Aptera, faster than Edison2, faster than everyone. And not a single cone out of place.

part two

knockout

As the component parts of all new machines may be said to be old . . . the mechanic should sit down among the levers, screws, wedges, wheels, etc. like a poet among the letters of the alphabet considering them as the exhibition of his thoughts; in which a new arrangement transmits a new idea to the world.

—Robert Fulton,
inventor of the steamboat

12

The Nobel Prize in Engineering

Kevin Smith started recording his ideas for inventions in middle school. He kept them in a series of notebooks he showed no one. When he filled up one notebook, he started another. On the back of the notebooks he recorded the 1-800 numbers of companies selling inventor's kits.

Sometimes he would invent something only to find that it had already been invented. "Medical adhesive," he wrote on the top of one page:

> I was working with pine branches today and was examining its ability to hold skin together like a glue & I thought, hey wouldn't it be great to have a glue instead of stiches [all sic] to hold cuts together for healing thus making smaller scars.
>
> Then Mike Balfour informed me that about 2 weeks ago the FDA OK'd just that thing.
>
> Close.

Kevin didn't segregate his ideas for inventions based on the weightiness of the topic. He would devote four haphazardly spelled pages to thoughts on superconductors, then move straight on to an idea for a cat toy or painting stilts. Between 1991 and 2007, Kevin wrote down ideas for building an open-ended ratchet wrench, a stick welder, bullet-deflecting material, cigarettes "with something besides tobacco leaves," Teflon-coated wall tilings, a submarine, a wind-powered plane, a photovoltaic cell, an Etch A Sketch–style chalkboard, a hard-plastic collapsible tent, a "macrowave" for flash-freezing food, a passive solar collector, a backyard astronomical observatory, a pulsed-ion propulsion drive for a starship, a magneto-dynamic levitation device, Ever Cold Beer Mugs, Ever Warm Coffee Mugs, and dozens of other contraptions both humble and grandiose. For example, in 1993, when Kevin was twenty-two, he wrote:

ALTERNATIVE ELECTRIC POWER

How about water powered generators that work off of the waves coming off of the sea. Sort of like windmills.

A weight would be sunk to the bottom with a cable attached to it. The cable would extend up to a floating platform and is connected to a type of piston or other form of generator. As the waves come in the platform is lifted and the piston or other apparatus is pulled over a distance against a constant preasure of a return spring or weighted cable and this mechanical work could either be directly utilized for a pump or made to generate electricity. These design plans and possible uses are in no way limited to the given information. The concept is a creation of Kevin Smith and all rights thereof are his to utilize as he deems fit.

When I asked a few engineer friends of mine about this idea, they said it would never work. Connecting a cable from the bottom of the sea to a piston on the surface is guaranteed to fail; tides and waves would thrash the cable around. But the basic idea of generating power from ocean waves is sound. Companies in New Zealand and Canada are currently pursuing the technology, using underwater turbines as well as devices to exploit the difference in heat between deep and shallow water.

Money was never the impetus for Kevin's ideas, although when he wrote down his idea for tobacco-less cigarettes he did note that in addition to reducing addiction there was the potential for "good profits." His prime motive was the joy of discovering either some new arrangement of things or some fundamental truth about the universe. The new arrangement or truth could involve a soda can or it could involve a photon. If you filled empty soda cans with sand and strung them together with rope, couldn't you make a very strong and flexible river dam? What does it mean that light is both particle and wave?

After he learned that the particle-wave duality had already been solved in the 1920s, by the founders of quantum mechanics, Kevin's thoughts increasingly focused on the riddle of gravity, the major unsolved problem in physics. Every so often he would scribble "Gravity" at the top of a blank page and ruminate for hundreds of words. And at the same time that Kevin wondered about gravity, he wondered if he was smart enough to be wondering about it. His journals exhibit a painful awareness that it's kind of odd for the son of a grocery-store manager to be scratching out diagrams for quantum-tunneling devices.

I showed Kevin's journals to a friend of mine, Alexandra Pechhold Zenker, a patent attorney who worked at the U.S. Patent and Trademark Office in Alexandria, Virginia, from 1999 to 2006. During that time she examined hundreds of patent applications. She also has a degree in civil engineering. Zenker told me that Kevin is "definitely a very creative guy who thinks outside the box," and that she found some of his ideas compelling. At the same time, they weren't properly fleshed out. The applications that once landed on her desk were accompanied by sketches, diagrams, evidence of testing, and pages of documentation. "What I found missing was, he jumps from idea to idea, and he knows it himself," she said.

Kevin was always looking for ways to tell if he was truly smart. He read biographies of Edison and Tesla to see if he shared any personal qualities with those great men. He took an IQ test and the Mensa test, averaged his scores, and got 156—a high number, but still just a number. He preferred to test himself by building things. And if he could do it in competition with other builders, all the better, because then he could measure himself against his peers.

The toothpick bridges came first, in high school. Then, as a freshman in college, Kevin saw a banner above a cafeteria that said ENGINEERS' WEEK. He walked inside, saw a bunch of teams building bridges out of Tinkertoys, and asked if he could enter. Someone handed him a box of toys, and in no time at all, Kevin threw together a bridge that held 125 pounds of barbell weights. The next closest team's bridge held only 15 pounds.

The engineers on that team turned out to be members of the school's Society of Automotive Engineers (SAE) club, and they were so impressed with Kevin that they invited him to help work on a solar car they were building. SAE designed and built hybrid and electric cars and raced them against cars made by other schools. These competitions were a big deal. Major automotive companies donated vehicles to the collegiate teams, who worked on their cars for years before a race. And members of the winning teams often got engineering jobs in Detroit.

At first, the SAE guys didn't know what to make of Kevin. He would tell them stories about working on cars in high school, stories almost too crazy to be believed, like the one about the time he needed to dispose of a rusting hulk of a car but didn't know whom to call to have it towed to the dump, so he simply hacksawed it apart, bit by bit, and dragged the bits to the curb over a period of months. *Here was a man who cut a car into pieces small enough to throw away in the trash.* He seemed to have inhuman staying power. Late at night, he would suddenly appear at the side of Walt Gorczowski, who would later serve as president of the SAE. Kevin's forehead would be moist, his pupils dilated, his faint mustache smeared with the chocolate chip cookie dough he liked to eat to keep himself awake. He'd open his palm and shove a small mock-up of a car or a robot or a bridge into Walt's face.

"Try to break it."

"Kev, I'm tired, I want to go to bed."

"It's *so strong*, though."

Kevin and Walt soon became best friends and roommates. In 1992, they entered the same Tinkertoy contest that Kevin had won as a freshman. Working as a team, they managed to make a Tinkertoy bridge that held 300 pounds of barbell weights. The plastic turned funny colors from the strain, but the bridge did not break. Then Kevin and Walt stood on top of it. The bridge made a sound like a log on a fire, but held.

"Do you want your trophy now or later?" the judge asked, rolling his eyes.

In another project, Kevin and Walt built a robot to win a King of the Hill contest. There was a plywood pyramid, 4 feet square and a foot and a half tall, with a flat top. The objective was to roll your robot to the top and fight off the other robots trying to do the same. Some competitors spent hundreds of hours building intricate mobile robots, giving them names like Eliminator and Devastator. Kevin and Walt called their robot Joe. Unlike the other robots, Joe didn't move or walk. It looked like one of those novelty toy birds that dips its beak into a glass of water over and over. At the bottom was a tripod. A 10-foot-tall steel post rose from the tripod. Attached to the post, a foot and a half off the ground, was a 3-foot pivot arm, and at the end of the arm dangled a steel plate with spikes. On the plate, Kevin and Walt had painted a smiley face.

Joe had taken about two hours and less than a hundred dollars to build. When it came time for the robots to battle, all Joe had to do was "fall" forward onto the hill. If another robot was on the hill, Joe's metal plate smashed it. If not, Joe blocked the way. Joe crushed most of the other robots into beer cans inside of ten minutes.

"Kevin didn't come up with all of the ideas or most of the ideas, but he was at the *center* of the ideas," recalls Walt. "He would be the cheerleader. Come on, guys, let's think of something. He would just throw ridiculous ideas out there, and they'd be honed into something better."

As graduation approached, the paths of Kevin and Walt began to diverge. Walt hunted for a job at an automotive company; he would eventually land at a large manufacturer of commercial trucks and truck engines. Kevin preferred to work for himself. Along with two friends, he started a company, Overload Engineering. Overload's first project was an artificially illuminated, fiber-optic Christmas tree—one that would ship with the lights already built in, sprinkled all throughout the tree. Kevin worked on the tree with his buddies. It was simple and clever; each of the many lights would consist of two beads of colored plastic atop a little fiber-optic lanyard that wrapped around the tree, and if you wanted to change the color of the lights, you merely replaced the beads. Kevin and his partners applied for a patent in 1994. The year after theirs was approved, in 1996, dozens of similar patents came out, filed by Christmas-tree manufacturers. It could have been a coincidence, but Kevin suspected that the manufacturers had copied his idea. He didn't have any money to sue to protect his patent. "You know how much money I saw from that?" he said. "Zeeeeeeooooooo!"

But Overload's failure didn't stop the flow of Kevin's ideas. He

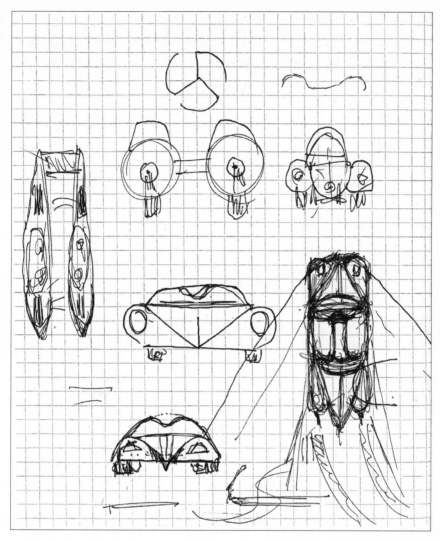

Kevin's first sketches of the car.

continued to fill his notebooks with descriptions of inventions even after he graduated from college, got a job at the Illinois EPA, bought the house in Divernon, and married Jen Danzinger. In his thirties now, and lacking a sounding board for his ideas, Kevin often tried to bounce them off his wife, with mixed results. As Jen came out of the bathroom, he would ambush her, aflame with an idea about string theory. While Jen walked the dogs, Kevin would air his latest idea about solar cells. He even chatted with her about inventions over instant messaging when she was at work. Kevin saved the transcript of one of their chats and stapled it into his journal:

> oddjobs42: hey, just thought of something
>
> oddjobs42: remember I said you could send messages through space/time if they didn't get all garbled like a badly burned cd?
>
> oddjobs42: well, I know how to do it if it's possible
>
> oddjobs42: simple morse code
>
> oddjobs42: a dimensional "pump" will be needed to send burst of energy into 1d space . . . the burst of energy . . . could be interpreted like the dots and dashes of morse code
>
> oddjobs42: ☺
>
> **Danz:** dimensional pump? . . .
>
> oddjobs42: does that make any sense at all?
>
> **Danz:** No but that's ok.

Jen wondered if there was a way to harness Kevin's energy and inventiveness to an actual paying job. One year, she subscribed to *Ode*, a magazine about green technology, published in the Netherlands. Every month, the magazine would make its way from watery, progressive Rotterdam to the pickup truck of their mail carrier in the heart of coal-fired Illinois. Jen thrilled to the stories about little European utopias, places in France and Sweden and Denmark where people with interesting ideas were getting paid to push the limits of the possible.

Jen borrowed a book about Denmark from the library. She showed it to Kevin and asked if he would ever consider moving to a place like that. He responded with a barrage of questions. What would happen to their pets? What if he couldn't find a job in Denmark? What about moving away from their families? That was the last time Jen brought up Denmark.

IT WAS a guy at work who first told him, in 2007. A car prize. Ten million bucks. The words "X Prize" rang a bell with Kevin. He remembered the space prize from three years earlier, remembered being captivated by the story of the maverick engineer named Rutan.

On a lunch break at the Illinois EPA, Kevin and his coworker started running some numbers. Was a 100-mile-per-gallon car even possible? They created a spreadsheet with several columns. WEIGHT, in pounds. DRIVETRAIN EFFICIENCY—the percentage of the car's fuel that is converted to forward motion. AERODYNAMIC DRAG. FRONTAL AREA. He plugged in various numbers. He saw that as long as he could keep the drag low and the frontal area down to about 1.5 square meters, about the size of a closet door, he could achieve 100 MPG with either an electric or a hybrid vehicle. It was possible.

Kevin began to sketch cars on a piece of notebook paper. What was the most aerodynamic shape he knew? Immediately he thought of *Star Wars*. In particular, his mind snapped to the Cloud City scenes from *The Empire Strikes Back*, in which merchants traveling in teardrop-shaped "cloud cars" slice through the sky. But the teardrop on the cloud car started with a point in front and got round in back. From an aerodynamic perspective, this was backwards; what you wanted was a rounded front that tapered gradually to a point in the back, preventing turbulence in the rear and bringing the air together in a gentle, controlled manner. What if you flipped it? Kevin imagined a reverse cloud car that seated two people, tandem-style, one in front and one in back. Then he imagined two such vehicles, side by side, squished together and welded down the middle. In theory, this would create a sedan big enough to fit four large humans, as required by the X Prize guidelines: a long, air-slick, tapered shape.

Kevin's sketches started to resemble a Volkswagen Beetle that had been pulled like a piece of taffy; there was also a hint of the original Porsche, the 356, which looked like an upside-down bathtub. For inspiration and reassurance, he looked at aerodynamic cars from the thirties, with their long tapered rears, like the tail of a teardrop, and their voluptuous fenders. He found a picture of the 1935 Delahaye 135M, a low-slung French luxury car with fenders that looked like torpedoes hanging from the belly of a plane. People had made cars like this once, for sound physical reasons, and drivers had bought them—maybe not the most radical designs, like the 1934 Chrysler Airflow, but more modestly streamlined and tapered models like the Lincoln Zephyr, Lincoln's bestselling car for years in the late thirties and early forties. Who was to say that the style couldn't come back?

Because he wasn't rich, Kevin knew he'd have to make his car out of the cheapest available materials, including parts from junkyards. So, along with two other guys at the EPA, he started spending his lunches in the parking lot, sneaking under various vehicles with a tape measure, trying to get a sense of which transmission might have the right dimensions for his purposes, and which front suspension, and which rear suspension. Not everything could be built from existing parts, though. Kevin didn't know of a chassis and body aerodynamic enough to get to 100 MPG.

A plan was forming. First, Kevin would build the chassis—the skeleton—from scratch, by hand. Then he would fill this handmade chassis with organs from other cars. Finally, he would create a custom aerodynamic body to wrap around the parts and the people.

The problem was, Kevin didn't feel comfortable with the idea of leading the team. That had never been his role back in college when he worked on experimental cars. Walt, his old roommate, had always been the leader. So Kevin called Walt and mentioned the X Prize. Maybe Walt would agree to take the reins.

Walt drove down from Chicago and met Kevin in Divernon. Out in the barn, Kevin explained his idea for the car. He talked about the spreadsheets and calculations he had done and asked Walt what he thought. Walt looked around the barn for a minute. Then he said that there was no conspiracy to keep gas mileage down. He spoke to the major automakers every day. He knew good people who worked at those companies. If a 100 MPG car were possible, the big automakers would have already made one.

Walt suggested that Kevin try a different hobby.

He saw the look on Kevin's face. Then he said, "Well, if anyone *could* do it, it would be you."

KEVIN DISAGREED with Walt about why the big automakers hadn't been able to produce more efficient cars. Walt thought the automakers were trying, and the task was just really hard. Kevin thought they weren't trying because they were too invested in old technologies. But it didn't matter what either of them believed; all that mattered was what Kevin could prove on the X Prize track. He thought back to the science books that Yoda had showed him, the ones full of stories about men who'd won Nobel Prizes for their discoveries in physics. There was no such thing as a Nobel Prize in Engineering. But if there were, Kevin believed, it would look a lot like the X Prize.

On August 31, 2007, he overnighted his application and a $1,000 registration check to the X Prize Foundation headquarters. Shortly after that, he sold his motorcycle. Then he and Jen went to the bank and took out a home equity loan on their paid-off house so that Kevin could afford to renovate the barn into a working garage.

This was a major break from domestic protocol. Jen and Kevin

had always kept separate bank accounts. Jen valued her independence too highly to abide another person having veto power over
her expenses, and vice versa. But they both knew that the Prize
was too big for them to treat as just another project. Before sending in the application, Kevin went to Jen and asked for her permis-
sion. "You won't see me for a couple years," he said, trying to talk
her out of it. But she encouraged him to enter. Jen could tell he'd
regret it if he didn't try. And she knew that Walt's criticism had
stung Kevin, making him feel like he had something to prove.

Over the course of two months, along with several friends,
Kevin refurbished the barn. After sweeping up all the mouse turds,
they poured a concrete floor and dug a trench to run power lines.
As soon as the concrete set, Kevin grabbed a piece of chalk and
knelt on the floor. He drew the future car in outline: 17.5 feet long
(about the length of a Lincoln Town Car), 70 inches wide at the
widest point (1 inch wider than a Toyota Prius), tapering severely
in the middle. On the rear wall, Kevin's friend Josh Spradlin sten-
ciled the name Kevin had chosen for their venture: Illuminati
Motor Works.

The name was a practical joker's wink to the wacko theory
about the Illuminati, a shadowy group of Bavarians who had sup-
posedly discovered, centuries ago, a limitless source of free energy
that was later suppressed by the oil companies. But the goal of Illuminati Motor Works wasn't to break the laws of physics. It was to
rub against them and find a new application. Below the company
name, Josh painted an appropriately occult design: an interlocking
triangle and circle, similar to the Masons' logo, with a single dis-
embodied eye floating in the center. In that spirit, they decided to
call the car Seven, a number that recurred in Masonic rituals.

Word spread quickly throughout the county that something
cool was happening in Divernon. The Springfield newspaper
sent a reporter to the workshop. The day after the article ran,
Josh heard footsteps approaching him in the Abraham Lincoln

Presidential Library, where he worked as a technician, scanning newspapers onto microfilm. Looking up, he recognized the close-cropped blond hair, red beard, and freckled complexion of Nate Knappenburger, who worked in another part of the library. Nate maintained the audiovisual kiosks and films. More important, he reanimated Abraham Lincoln at eight-thirty every morning. The museum's pride and joy was its holographic theater, in which a live actor traded lines with a series of 3-D holographic movies of the sixteenth president. Before working at the Library, Nate had maintained F-16 fighter jets for the U.S. Air Force, until he failed a physical ("I got fat," he recalls) and was honorably discharged. After that he had studied electronics at ITT Tech.

Nate held up the newspaper to Josh, his catcher's mitt of a hand covering up the article.

"I want to do this," he said.

Over at the Illinois EPA, another person was expressing interest in the project—a sixty-two-year-old Air Force veteran named George Kennedy. George wrote biodiesel permits, helping farmers turn their soybeans into fuel. The work had made George skeptical of biodiesel's potential to cut down on the use of oil. According to his calculations, the entire soybean crop of Illinois would have to be cultivated and processed in order to reduce the state's consumption of oil by even a fraction of a percent. Oil was still king. And oil, it seemed to George, had more than a little to do with the fact that the Illinois National Guard had deployed his twenty-five-year-old son to Iraq.

George approached Kevin's cubicle.

"You know, somebody has to do something," George said, placing a $20 bill on Kevin's desk. "Donation," he muttered, and walked away before Kevin could shout, "Wow, thanks George!"

A week later, George returned. "You know, $20 won't take you very far," he said, and handed Kevin $50.

"Gee, thanks George!"

The following week, when George stopped by, he was holding a laptop.

"My wife says I need a hobby," George told Kevin. "I used to do CAD. You said you needed help with CAD?"

It turned out that George held a master's degree in mechanical engineering. The last time he'd worked in computers was thirty years ago. Still, he was willing to relearn his old skills in order to build a detailed 3-D computer model of Kevin's car, which Kevin knew would be hugely useful during construction.

A sixty-two-year-old computer guy?

"Awesome, George!"

THEY DECIDED to make a plug-in electric car. It seemed like a no-brainer. An electric motor was 90 percent efficient on a bad day, as opposed to 25 percent or less for a gasoline engine. Going electric was simply the cheapest and simplest way to make a superefficient car—far cheaper than taking weight out of the car by using carbon fiber or aluminum, or by inventing an entirely new automotive architecture like Edison2 was doing. If Illuminati built an elegant electric drivetrain and coupled it with a sleek body, the car could use heavy junkyard parts and still be competitive.

Now they had to find a motor and batteries they could afford, which turned out to be a challenge. There aren't a lot of companies that sell these items for electric cars, for the obvious reason that there aren't a lot of electric cars. You can't go to Pep Boys and pluck a motor and battery pack off the shelf.

At first, Kevin thought he could get around the problem by asking battery manufacturers to sponsor him—to give him free or reduced-price equipment in exchange for a logo on the car—but he didn't get very far. When he called LG Electronics, a South Korean company, no one called him back, and when he reached A123, a small battery company in Massachusetts, a polite woman explained that A123 was interested in working with him only if he planned to manufacture thousands of cars. (Cofounded by an MIT scientist and backstopped by a $249 million Department of Energy grant, A123 would later become an official sponsor of Aptera Motors.) About the only business that seemed excited about Kevin's sponsorship pitch was a hardware store in Springfield, Noonan True Value, which agreed to sell bolts and paint to him at cost.

He ended up buying his electric motor from a Russian guy in Portland, Oregon. Josh had stumbled across him on Google. The guy owned a small company, Metric Mind, that provided electric-vehicle components to hobbyists interested in converting their daily drivers to EVs. The motor Kevin bought—designed in Switzerland, manufactured in Denmark, and commonly used in European trucks—was the most affordable one he could find that was large enough to power the car. It packed 200 horsepower, compared with 149 horsepower for the motor that would eventually power the Chevy Volt.

He found a Chinese company that sold the kind of batteries he wanted, under the brand name Thunder Sky. Kevin bought 96 small Thunder Sky batteries to string together into a proper battery pack. The batteries used a specific type of lithium-ion chemistry called lithium iron phosphate. The chemical bonds in lithium iron phosphate made them less likely than laptop or cell phone batteries to catch on fire if they overheated; Kevin thought they'd give the car more reliability. All told, the batteries would provide the motor with 33 kilowatt hours of energy—9 more kilowatt hours than the pack that Nissan would later put in its plug-in electric car, the Leaf.

After months of work, the car began to take shape. To make Seven's structural curves, Kevin and Josh melted steel tubes in the woodstove. To determine the form of the curves, Kevin relied on a chart he'd stumbled across in *The Leading Edge*, his aerodynamics textbook. The chart—listing a series of basic shapes and their coefficients of drag, or Cd—amounted to the ultimate aerodynamics scouting report. It laid everything out in black-and-white, from the bad to the good, starting with a brick.

The coefficient of drag of a brick is 2. That's high, and high is bad. The Empire State Building, according to the chart, is slightly better, with a Cd of 1.5; the structure tapers near the top. A cone is 0.5. A teardrop clocks in at just 0.05 if the fat part of the teardrop is facing forward—4 times sleeker than the lowest-drag production car ever made (the General Motors EV-1), 10 times sleeker than a cone, and 30 times sleeker than the Empire State. Another section of the chart talked about sea mammals. A dolphin in water has a Cd of 0.004 based on the total wetted area of its skin. A sea lion by this same measure is 0.0041; a penguin is 0.0044. A human swimmer comes in at 0.035, about ten times clumsier than a penguin, which makes sense: We didn't evolve in the water, and penguins did. Nature is an aerodynamicist.

Kevin looked at the chart and realized he didn't need to know everything about aerodynamics: "I'm not making the Space Shuttle here." All he needed was a few rules of thumb. The chart reassured him, because it confirmed that the cloud-car concept, based on the teardrop shape, was fundamentally sound. (Kevin once told me that if I wanted to radically shorten this section of the book, I could write it in two words: "Kevin guessed.")

Once Kevin had welded the curves together to form Seven's steel-tube frame, Kevin's father, Nick, and his brother, John, cut foam-panel sheeting, plunked it down between the spots on the frame, and covered it with two layers of fiberglass and epoxy resin. Then Kevin and Josh sprayed the whole thing with lightweight fire-retardant foam the color and consistency of the head on a pint of Guinness. Now it was starting to look like a real car. Eventually, they added some more subtle touches, including headlight covers from a thirties-vintage Ford. They put in a stereo sound system, which was required by the X Prize rules; it had an FM radio and could play MP3s.

In June 2009, two years into the build, Illuminati had to submit a large stack of paperwork to the Prize. One element was a business plan. According to the Prize, it had to be "a credible plan to manufacture, sell, and service 10,000 vehicles (or conversions) per year by 2014. The plan must show that the national fuel infrastructure will support the vehicles, especially if any non-standard fuels or fueling-methods are to be used." Kevin didn't spend much time on the business plan. He thought the exercise was silly, and it showed in the document, which had the coerced, empty optimism of a middle-school book report. Written with George, it called for an Illuminati production facility in Springfield that would cost $40 million up front and $40 million over four years—$80 million in a town whose government survived on a shrinking yearly budget of $105 million, in a state with a $10.2 billion budget gap.

Kevin took the other parts of the paperwork submission more seriously. He had to show that his car would meet all Prize rules as well as conform to a subset of the exquisitely detailed Federal Motor Vehicle Safety Standards (FMVSS) that require cars to be built in such a way as to protect their occupants from a hundred varieties of disaster. A lot of the other teams were tearing their hair out over the safety stuff; the FMVSS regulations on seat belts alone ran to twenty pages. This part should have been especially hard for Illuminati, since they were making a car from scratch instead of starting with one already designed to pass federal muster. But because Kevin and his teammates were government bureaucrats, experienced at reading long documents and extracting the vital nuggets, they whipped out the safety section in a day.

The more challenging problem was estimating the car's likely performance. To do that, Kevin had to plug a series of numbers into a spreadsheet provided by the Prize: the power of the car's motor or engine, the total weight of the car, the rolling resistance of its tires. One of the required numbers was the car's coefficient of drag. Kevin wasn't sure how to calculate it. He took Friday off work, sat down on his couch with *The Leading Edge,* and started working through various equations. He assumed that the car's curves weren't perfect, that the taper on the tail wasn't maximally smooth, because, well, no curve is perfect; even a straight line isn't straight. Stretch it out long enough and it curves. The universe curves. It took Kevin four days to figure out how to solve for drag, but eventually a number did emerge: .165. He entered that into the spreadsheet, and a set of numbers popped out. Ridiculous numbers.

Okay, Kevin thought, *I must be doing something wrong.*

He checked the math again and got the same answers.

Top speed, 225 miles per hour. That was the theoretical limit of the electric motor. In reality, Kevin wouldn't be able to top 130, because the car's tires couldn't handle higher speeds.

Fuel economy, 579 MPGe.

Range on a single battery charge, 500 miles.

A 500-mile range? "The X Prize isn't gonna like this," Kevin told George. So he changed the variables. On paper, he crapped up his car. Instead of assuming a 90-percent-efficient electric drivetrain, he assumed the efficiency of a 1970s Chevy Impala. Then he doubled the drag, going from a sleek .165 to a .36, the same as a Volkswagen Beetle. "We know we're better than point-three-six. Point-four is a minivan. Point-three-six is like a small truck. My wife's Gremlin is probably a point-three." He and George ran the numbers again, and this time, the fuel economy came out to 260 MPGe, and the range was 240 miles. Still implausibly high. So on his technical submission to the Prize, Kevin attached a note, showing all of his math and spelling out his assumptions. The note said, essentially, "Please tell us if we're crazy."

A few days later, the X Prize Foundation e-mailed Kevin. His submission had been accepted without comment.

13

Yesterday Today and Forever

The original American inventor was cold and hungry. He came on a boat to a new raw place: Jamestown, 1607. Trees, creeks, not much food. He'd brought some armor to defend against the Spanish, but the Spanish didn't show, so he chopped it into cookware, according to *They Made America* by Harold Evans.

Years passed. More people came on boats. They spread through the colonies, improvising new tools. They made a better axe, a better stove. The stove guy, Benjamin Franklin, also had some ideas about government and democracy. He and his friends fought a war to make those ideas real.

Even as they fought, they were thinking about how to spawn more inventions. For the nation to survive, they realized, it had to grow; to grow, it had to invent. Thomas Jefferson designed America's first patent system, despite his instinctual dislike of patents; an accomplished inventor himself, he believed all people should be able to enjoy the fruits of invention, not just the creator. "He who receives an idea from me, receives instruction himself without lessening mine," Jefferson once wrote in a letter, "as he who lights his taper at mine, receives light without darkening me." In Jefferson's scheme, enshrined in the Constitution, any American could apply to patent an invention as long as it was useful and original. Frivolous patents would be rejected by educated "examiners." Jefferson served as examiner on the first U.S. patent, awarded in 1790 to a Vermont man for a new method of making an industrial chemical called potash.

Within two decades, the government was issuing more than two hundred patents each year. But patents alone didn't define the American inventor. The system only corralled a small portion of the larger creative energy that drove the growth of the nation. Men and women bashed into new territories on vehicles of their own making. They completed large engineering projects that made the travel easier. In the 1820s, thousands of yeoman farmers dug the Erie Canal, a 360-mile gash across the malarial swamp known as New York. A formal engineering profession didn't yet exist in America; before this, engineers had been imported from England and France. The farmers building the canal were supervised by an itinerant elementary-school math teacher, a tree-clearing axman, and a country surveyor. "There was not a man of any reputation for science employed in the work," novelist James Fenimore Cooper later wrote. "But the utmost practical knowledge of men and of things was manifested in the whole of the affair."

This was the crucial trait: practicality. The American inventor was an ordinary person able to make his way in the world by transforming the materials at hand into useful things. Not a genius but *ingenious*. Genius was royal, remote, but ingenuity was democratic; only a few people could be geniuses, but anyone could show ingenuity. Even the Americans commonly hailed as geniuses often tended to be something humbler. Thomas Edison, who founded his famous research laboratory in Menlo Park, New Jersey, in 1876, wasn't the smartest electrical engineer of his day; Nikola Tesla was.

But Edison won the glory, and Tesla died broke, on the thirty-third floor of the New Yorker Hotel, because "The Wizard of Menlo Park" was more ingenious: canny, self-promoting, commercial, and, above all, practical.

Ingenuity continued to drive America's inventive life even into the twentieth century, as corporations became enormous enough to support large labs of their own, and the age of the individual inventor gave way to the age of industrial science. The company that exemplified this transition was Bell Telephone Laboratories, originally founded by Alexander Graham Bell, inventor of the telephone, and later subsumed into AT&T, the telephone monopoly. By the tail end of the Great Depression, Bell Labs had expanded into a bustling operation of five thousand scientists, engineers, and assistants, most working in a ten-story building in New York. "Some of them resemble prosperous college department heads; some look like bank vice presidents," according to a 1931 *New Yorker* piece about Bell Labs. "None of them assume the airs of genius burning in secret."

In 1941, Bell Labs moved to New Jersey, to a new building designed to force people to collaborate across disciplines—an ingenious concoction in its own right. Some of the hallways "were designed to be so long that to look down their length was to see the end disappear at a vanishing point," writes Bell Labs historian Jon Gertner. "Traveling the hall's length without encountering a number of acquaintances, problems, diversions and ideas was almost impossible. A physicist on his way to lunch in the cafeteria was like a magnet rolling past iron filings." It's hard to fathom how wildly this approach worked. In 1947, Bell Labs researchers invented the transistor, the technology now at the heart of every computer and electronic device. They invented the Unix operating system, various flavors of which now run the computers that run the Internet. They invented the fax machine and the first long-distance TV transmission, the solar battery and the transatlantic telephone cable. They invented and flicked on the world's first lasers, light-emitting diodes (LEDs), and cellular telephone signals.

Other invention factories arose in America before and after World War II, bunkers where scientists and engineers worked in relative freedom, insulated from the pressures of the market by corporate might or military necessity. There was NASA: Mercury, Gemini, Apollo, the Moon Shot, the great slide-rule yawp of the postwar engineers. "The mechanical guys, the aerodynamicists, they were the kings," Oliver Kuttner says. "The best were building the Space Shuttle, and building ICBMs, and F14s, and shit like that, you know?" There was a company called RCA Labs, which got its start researching radar techniques and later invented the standard for color TV. There were scattered labs where a new breed of engineer studied how to make computers faster, cheaper, and more connected. In 1969, a handful of these guys built the forerunner of the Internet, ARPANET, on behalf of the military's R&D wing, the Advanced Research Projects Agency (ARPA, later renamed DARPA). A little later, in Palo Alto, California, engineers working for Xerox invented the graphical user interface and the computer mouse, concepts whose commercial potential was lost on Xerox but not on an acid-dropping twenty-four-year-old named Steve Jobs, who visited Xerox in 1979, hurried back to his little company, Apple Computer, and told a colleague he had to make a mouse.

Then something sad happened. In the eighties and nineties, in the name of budget savings and short-term profits, the engines of American invention forfeited their powers. NASA ramped down, struggling to find a mission that made sense in the absence of the Soviet Union. Ronald Reagan and George H. W. Bush pushed defense spending at the expense of science and technology; DARPA veered into making military machinery. Corporations outsourced jobs and factories to other countries, making it harder to quickly scale up ideas in America. Bell Labs was spun off into Lucent Technologies, which later sold itself to a French company;

by 2009, only one thousand people worked at Bell Labs, down from thirty thousand in the early 2000s. Apple now made its computers and smartphones in China, in a factory where managers had to install nets to catch the people trying to commit suicide by flinging themselves off the side of the building.

In theory, small inventors should have been able to fill the void left by the big companies. But over the decades, the U.S. patent system had corroded. The Patent Office was issuing more patents than ever—90,365 in 1990, 157,494 in 2000, 219,614 in 2010—but according to an investigation by the *Washington Monthly*, as many as half were faulty. Companies and their patent attorneys had learned how to prey on low-paid patent clerks, flooding them with applications designed to protect corporate turf, not useful and original ideas. Junk patents proliferated, crowding out the little guy.

As a result, by the late 2000s, it seemed like a sucker's bet to try to make a living as an inventor in the classic sense, by creating useful and original things. In the popular culture, the word "invention" was synonymous with getting rich quick; a number 1 bestselling book, *The 4-Hour Workweek*, offered advice on joining "the new rich" by creating a "virtual architecture" to sell products that could be invented at minimal cost. "Innovative gadgets or devices are great but often require special tooling," author Tim Ferriss wrote, "which makes the manufacturing start-up costs too expensive to meet our criteria." (Ferriss got his start by conceiving and selling a nutrient supplement called BrainQUICKEN.) Meanwhile, the country's most famous inventors were inventing things of dubious merit, generating enormous wealth for a few by hawking gadgets to the many. In the San Francisco Bay Area, as America's coal-fired power plants continued to soak the atmosphere with gunk, as dysfunction snarled Congress and the roads and bridges chipped and cracked, as twelve million searched in vain for jobs and the economies of entire towns ran on food stamps, the best and brightest trilled about the awesomeness of their smartphone apps. Twitter,

Facebook, Instagram, Angry Birds, Summly, Wavii: software to entertain, encapsulate, package, distract. Silicon Valley: a place that has made many useful things and created enormous wealth and transformed the way we live and where many are now working to build a virtual social layer atop the real corroding world.

THE X Prize Foundation had many connections to Silicon Valley. Its leader, Peter Diamandis, traveled there all the time, flying north from X Prize headquarters outside of Los Angeles, where he lived. The Foundation also shared two core values with the big firms and venture capitalists: a faith in the power of private markets and a utopian view of technology's potential. But in at least one important way, the Foundation was the inverse of a Silicon Valley institution: It cared exclusively about hard, real-world problems. Disease, poverty, energy, climate change, transportation.

"I'm fascinated about the future of transportation across the board, and I see this multidimensional evolution," Diamandis told me one day between stages of the Prize. We were in the Bay Area, walking to lunch under a blue sky, across what felt like the quad of a university, with low buildings all around. It was actually a NASA campus—the NASA Ames Research Center, eight miles from the headquarters of Apple, in Cupertino. Diamandis gestured toward a dude in shorts and a T-shirt walking in front of us. "That's an astronaut," he said.

I'd come to see him because I wanted to understand how the Automotive X Prize fit into the Foundation's broader vision. Diamandis took long, ponderous strides, wheeling a black airline bag behind him. Noticing that I was struggling to hold my voice recorder close to him as he walked, he offered to hold it for me. He raised it to his lips and spoke into it as if it were a microphone at a podium.

"I think humans are the worst control systems to put in cars," he said. In the future, cars would drive themselves—faster, more safely, and more efficiently than humans ever could. Engineers at Google, Carnegie Mellon University, and Stanford University were already working on robotic car prototypes, and he hoped to develop a Robotic Car X Prize, in which robotic cars and human drivers would race one another through obstacle courses.

Beyond cars, he said, the Foundation was looking at electric aircraft, as well as a little Osprey-like helicopter he called Segway in the Sky. "Hop into a vehicle, push a button, and it autonomously takes you from one point to another point, say, ten miles away." He was also exploring new ways of going into space, "but with the recognition that the future of transportation may not be physical atoms. So how do we look at bringing computer interfaces, and look at how we transport our consciousness, our personalities, our feelings, our thoughts, our reactions, versus transporting ourselves physically?"

How do we transport our consciousnesses, our feelings, our thoughts? In three hundred words, Diamandis had gone from the car to a concept known as the Singularity, first proposed by his friend, the prominent inventor and futurist Ray Kurzweil. The Singularity is the point at which machines become smarter than humans. When this happens, the argument goes, a new breed of intelligent computer will be able to interface with the human brain, dissolving the distinction between man and machine and allowing people to essentially live forever. Diamandis had told reporters that he considered the idea of the Singularity to be "completely obvious"; computer processing speed has long increased at an exponential rate, and if it continues to do so, the Singularity is the logical result. He recently told the *New York Times* that he plans to live to be seven hundred years old.

We arrived at a NASA cafeteria and took a seat at an outdoor table beneath a sun umbrella. Diamandis was dressed all in black. Just back from a ten-day trip to Morocco, he was still wearing his shirt in the local style, with the top two buttons undone, revealing black chest hair. Spotting some of his students at the next table, he stood up, excused himself, hugged each student, and returned to the table, beaming.

In 2008, with NASA's support, Diamandis had started a school here, called Singularity University, dedicated to training future leaders in physics and engineering and to discovering what he calls "10^9+" technologies—solutions that will affect more than a billion people worldwide. The man who was once so critical of NASA's slowness and bloat was now being embraced by the agency. After lunch, he was scheduled to speak at a Singularity University conference attended by NASA's chief technologist, the rocket scientist Bobby Braun.

Diamandis set his smartphone on the table and glanced at it every thirty seconds, sometimes picking it up mid-sentence and peering at the screen, apologizing each time. When he focused on me, he was friendly and present and enormously likable. I asked him what he meant by "transporting our consciousnesses." He said, "We are, as a species, becoming more and more connected." At first, he said, we connected only with people in our tribe; then we were able to fly to another country; then connected instantly to anyone on the Internet. The next iteration would be "a meta intelligence—interconnectedness of all the human minds."

He mentioned the Borg, the aliens from *Star Trek* who exist as a single connected mind and whip through space in giant death cubes assimilating other civilizations into their hive consciousness. "Ultimately *Star Trek* gave the Borg a bad name," he said. "But if you imagine a kinder, gentler Borg." He smiled sheepishly, aware of how ridiculous this might sound to a person who hadn't yet accepted the inevitability of the Singularity, who hadn't yet crossed that bright metaphysical border. Then he said, with feeling, "It's not about losing individuality, it's about gaining intimacy with a larger number of people." He leaned forward. "So imagine you have the ability to plug in. You're standing here. We're having a conversation. And you plug into this meta-intelligence. And all of a sudden *you know anything you want to know*. You have the ability to see and to feel and know the knowledge of somebody on the other side of the planet, to experience the sun rising in Japan or a person's elation as they're climbing Mount Everest. It is an expansion of human consciousness to such a level that I think, once you've plugged in and experienced this, to unplug would be to feel so lonely."

It's easy to make fun of the guy from this side of the border, looking across. But Diamandis has a gift for making the lush new land seem reachable through a series of small, manageable steps. I asked him about his desire to live to seven hundred. He laughed. "Let's talk about that." Evolution has selected for humans to reach reproductive age, not to age gracefully; medicine has already tripled the human life span, from twenty-five to seventy-five; cell death is a product of genetic processes that can be understood and potentially reversed as computers that analyze DNA become more powerful. "Some of the oldest sea reptiles live to seven hundred years old, and I said if they can, why can't I?" he concluded.

This is why Diamandis is interesting: He blends the visionary tradition of science fiction with a deep practicality, coupled with a bedrock belief in the practical abilities of people he's never met. To prove that the faraway outcomes he cares about aren't as distant as they feel, he created an ingenious organization whose mission is to harvest the ingenuity of others. Diamandis isn't recruiting experts and elites to talk about problems. He's asking ordinary people to build things. The Foundation is one man's attempt to grasp at the future by resurrecting an idea of invention from the past.

There's also a more intimate way of looking at it. Diamandis isn't a guy driven by money or even by curiosity. What he really wants, at root, is instant access to the deepest and most meaningful experiences. This is clear when he describes the connected consciousness, and also when he talks about the teams that enter his prizes. They don't do it to get rich, he told me. "They do it because *it's why they've been put on this planet*. It's personal significance. It's wanting to know their life meant something. And part of my goal is to give people these targets that allow them to galvanize their time, their money, their friends, their resources. Sometimes people just need a target to shoot for."

In other words, Diamandis has created meaning in his own life by helping others find meaning in theirs. This is my sense, anyway, after absorbing his high-altitude view of things, where all is silent and peaceful, and the cars aren't even dots.

THERE ARE two kinds of people interested in the future of cars. One wants to make a better car. The other wants to get *beyond cars*: to create a world where the car is more tool than totem, a diminished part of a broader, smarter transportation system that encourages people to walk, bike, ride trains, and share cars instead of own them. A world angled toward debt-crushed young Americans, who aren't buying cars in nearly the numbers they used to, because who needs to own a $20,000 machine and shell out for gas when you can hang with your friends by smartphone or time-share a Zipcar?

Diamandis is definitely a beyond-car guy. He cares about cars to the extent that traditional ones impede the dawning of the science-fictiony world he wants to live in. But the people he enticed to compete in his auto prize are the opposite type. They're better-car people. They don't see the car as an inherent obstacle to human evolution. They still believe there's something glorious about the fact that you can hop in a car in Philly and point it south and be in Florida thirteen hours later. "The automobile, it's what made us who we are," Kevin Smith told me once. "How can we keep our freedom machine?" The Prize, to them, is about *saving* the automobile. It's about revealing the automobile's ultimate and ideal physical form—finishing, at long last, the job the big companies abandoned.

The tension between the beyond-car and better-car mind-sets was evident when the cars appeared on the track for the first time, at the Shakedown event in Michigan. Diamandis didn't visit the track for the first several days of tests, and when he finally arrived, he paced up and down Pit Lane, talking on his phone, looking distracted, out of place, a space kid surrounded by asphalt and mud.

All around him, cars were breaking. They broke on the oval and on the durability track. They broke doing easy laps and hard stops. They broke while accelerating and cruising and slaloming through the cones. Aptera needed forty tries to pass the Moose Test; two of Edison2's overstressed engines croaked in a single day. ("Blew up two," Oliver Kuttner told me with a wave of his hand, as if losing two engines was no big deal—an actual piece of luck, really, because it gave him a chance to show how *easy* it was to rebuild such a small engine.) During the durability test, West Philly's Focus hybrid briefly caught on fire; it turned out that, before the run, Simon Hauger had simply forgotten to tighten the oil cap, and freely sloshing oil had pooled and caught a spark. At the end of a day of Shakedown reporting, I was often relieved to get in my ratty 2002 Honda, because I knew it would start the first time I turned the key. A car I'd always viewed as an embarrassment began to seem like a miracle.

This was the basic paradox of the Prize: To get beyond the car, to a smooth and sparkling future of personal helos and space planes and connected consciousnesses, you first had to make a better car; and yet to make a better car, you had to test new electrical and mechanical systems, to debug and break and shatter, which created the sorts of visuals that made the future seem very far away.

Another paradox: As Diamandis put it in an antic speech to the teams over dinner one night, the Prize was about "changing the industry and really changing the entire *paradigm*, the perception, of what the public believes is possible. That you don't have to choose between beautiful, affordable, safe, and fast, and by the way it gets over a hundred miles per gallon equivalent. You can have it all." Yet to change public perception about cars, you had to first prove that new kinds of cars were possible, and the easiest way to get the proof was to test the cars in a controlled environment away from the public.

The Foundation had brought the Prize here to Michigan for sensible reasons. After realizing that a cross-country race would be too logistically hairy, organizers had settled on the Speedway because it would let them perform 95 percent of their tests in one spot. They rationalized the decision by invoking history. Michigan, birthplace of the automotive industry! As Eric Cahill, executive director of the auto prize, once put it to me, "If there's any better message than a cross-country race, it's: Here is the X Prize reinvigorating an industry in crisis, in the epicenter of the auto industry. The Rust Belt, right?"

Except when you actually got to the Speedway, it didn't ring any historical chimes. It could have been in Colorado or North Dakota or Vermont. It was just a racetrack in a field somewhere. Once you were inside the gates, you couldn't see anything except the two-mile loop of asphalt and the garages and the prototypes with hulls like licked candy weaving through traffic cones at relatively low speed and the electric cars clicking under a dome of blue sky. Shakedown ended up feeling like some kind of robot decathalon performed in an abandoned coliseum—NASCAR without the crowd or the danger. At night the teams left their prototypes in the white charging tent with the soaring big-top peaks and piled into Hondas and Fords and drove out through the Speedway's concrete exit tunnel, passing the road sign that said JESUS CHRIST THE SAME YESTERDAY TODAY AND FOREVER on their way to the Super8 or the Days Inn or the Travelodge. For a Promethean quest to relight the torch of American ingenuity it was about as glamorous as a continuing education class for country lawyers.

The lack of spectacle didn't do the Foundation any favors with the media. A handful of journalists did make the journey to Shakedown—writers from tech and green-energy websites, as well as a crew from the Discovery Channel, which was filming a documentary—and the reports they produced were full of geeky enthusiasm, if not evangelical zeal. Wrote *Popular Mechanics* in a typical piece, "Coming in from around the country on a ragtag fleet of haulers, [the teams] all had wide-eyed optimism on their side, even if they hadn't yet found time to actually paint their machines or otherwise give them the final gloss of a true production item." But the event was a tricky thing to cover. The auto prize wasn't simple and binary like the space prize. Space is *up there*. The plane goes up, the plane comes down. Did it crack 100 kilometers or not? Are the occupants still alive? Here at the Speedway, there was so much more to keep track of: not one clear test but a confusing range of them, involving cars in three different divisions, each with different rules.

"Any time that we proposed to write about it, we had to put in about five grafs of explanation about how it worked," recalls John Voelcker of Green Car Reports. His site ended up barely covering the Prize at all.

It was also hard to tell who might win, even after Shakedown. The event was too scattered to change people's preexisting opinions about which were the best cars. True, reporters thrashed Aptera on the blogs, posting screen grabs from a video of the door flying open during the Moose Test—"hardly ready for primetime," Green Car Reports wrote of the 2e, adding that Aptera was risking "corporate suicide" by testing it in public—but despite the car's troubles, its slick appearance carried the day. In an updated set of rankings, *Popular Mechanics* listed Aptera as the favorite to win "the main stash," at 3-to-2 odds: "It looks ready for production and is beautifully finished; a true consumer product." (Confusingly, the magazine had mixed up its stashes; Aptera wasn't in the running for the $5 million mainstream prize, only one of the $2.5 million prizes.)

Edison2 attracted far less publicity than Aptera. To the extent that auto writers noticed the Very Light Car, they seemed to view it as a less polished version of the Aptera 2e—a "poor man's Aptera," as AutoblogGreen put it. *Popular Mechanics* gave Oliver and his men 4-to-1 odds, recognizing that the Very Light Car was "a machine optimized for the competition" while raising doubts about the marketability of a vehicle that was "spartan in the extreme": "Would buyers accept that? Or feel comfortable in a car that seems so vulnerable in a world still filled with big trucks?"

Meanwhile, the press flocked to the teams with smaller budgets, though not because they respected them or thought they could win. Reporters took a particular liking to Illuminati Motor Works; Kevin was friendly and energetic and could grin into a camera and talk about the car for half an hour without stopping, and the fact that his wife was a team member provided a human-interest hook for TV crews, although their insatiable interest in the husband-and-wife angle made Jen in her shyness want to vomit and made the rest of the guys on the team feel ignored. "If the Illuminati's chances are slim, it's not for lack of audacity," *Popular Mechanics* wrote, laying 100-to-1 odds on an Illuminati win.

The magazine didn't list West Philly in its top contenders, but Simon Hauger and his students probably got more coverage than any of the teams, their rags-to-riches tale told and retold by reporters of all kinds. "From the grit and grime of a high school auto shop," wrote *Brass*, a magazine for young adults, "one group of students from West Philadelphia is challenging the expectations of a generation."

It's probably lucky for the Foundation that more people didn't pay attention to Shakedown, because it revealed a troubling omen that the design of the competition might be flawed. Two disreputable penny-stock companies had snuck into the field: Zap, from California, and Li-Ion Motors, a company founded to mine copper and gold in Canada but based in Las Vegas. Zap and Li-Ion were both notorious for promising cars that never actually got made. They would build prototypes, raise money from investors to go into production, then neglect to actually produce anything. According to a devastating 2008 profile of Zap in *Wired*, "investment firms around the country have become cautious about financing electric vehicles after being repeatedly misled by one of the industry's most visible companies." Now Zap was here at the X Prize, competing with a three-wheeled electric prototype it called the Alias. (They had 4-to-1 odds according to *Popular Mechanics*, which apparently hadn't caught on to what the rest of the industry knew, calling the Alias "a solid effort from a proven electric-car builder.") And Li-Ion had entered the Wave II, an electric car with an aerodynamic, lime-green body. ("A dark horse contender" with 30-to-1 odds.) It resembled a well-rubbed bar of Irish Spring.

What did it say about the Prize that Zap and Li-Ion could get this far? Didn't it suggest that the mileage target had been set too low at 100 MPGe, letting too many teams into the fold, including some that were subverting the Prize's true intent? "God, half the field was not credible," says Chelsea Sexton, the electric-vehicle advocate and original Prize director. How was it possible, for instance, that Edison2, a team using internal combustion engines, seemed to have a real chance? If you could win an efficiency contest by burning gasoline and ethanol, what the hell was the point?

And what about the car-buying masses? Where were their voices? Would anyone out there in America want to drive an Aptera 2e or a Very Light Car to work every day? Would anyone take their kids to school in an Illuminati Seven? Were these the cars the public wanted?

The way the Prize had been set up, it was impossible to know. Right now, anyway, officials couldn't afford to focus on the public.

They didn't have time to sell the sizzle on the steak when they were busy making sure the oven didn't burst into flames.

I DIDN'T understand how overwhelmed Prize officials were until I visited the Foundation's headquarters after Shakedown. The building, in Playa Vista, was easily identifiable from the street by the distinctive swooping X (one straight stroke, one curling) on the front door. Inside, space tackle sprawled through hallways and hung on walls: a watercolor of a cat in a spacesuit; a bookcase full of vintage sci-fi magazines; a road sign that said MARS—**35,000,000 MILES**.

A senior staffer named Eileen Bartholomew brought me into an office and started drawing boxes and circles on a whiteboard with a purple dry-erase marker. A slender woman with reddish-brown hair and a poised, fluid way of talking, she said she'd spent fifteen years as a consultant to biotech and pharmaceutical companies before joining the Foundation, and that working here had expanded her sense of the possible. "I don't see the world in those squares anymore," she said. "I see the squares, but I also see the holographic image." She filled the shapes on the whiteboard with words and phrases like *initiation, feasibility, ideation*, and *nuclei of innovation*. She said the original prize, the space prize, had been a seat-of-the-pants improvisation, and since then, the Foundation had made strides toward mechanizing the process: "We're trying to bring some science to the art."

The only active prizes at the time were the auto prize and the $30 million Google Lunar X Prize, which required teams to land a robot on the moon, drive it 500 meters, and beam information back to Earth. But the Foundation would soon announce its next prize, a $1.4 million Oil Cleanup X Prize, prompted by the disastrous explosion of the Deepwater Horizon rig—a prize eventually won by an Illinois team that improved oil-skimming rates by a factor of three. And the fifty employees here were considering at least a dozen other prizes, in fields from bionics to plastics to medical diagnostics.

Bartholomew said they'd developed a set of criteria to help them choose whether to launch a prize or take a pass: Was it both *achievable and audacious* ("hard enough to be a challenge but achievable enough to inspire")? Was it *measurable*? ("We don't want anybody not to understand why somebody won.") Did it have *telegenicity*—made-for-TV moments that would maximize and justify a sponsor's investment? Did the prize address a *market failure*—an instance where the flow of capital had been routed around an important problem? Finally, was the prize *operable*? Could you execute it? "Every prize, when we launch, is like a miniature business," Bartholomew said. "It has a life of three to eight years. It has a budget. It has a staff."

When I spoke with Eric Cahill in his windowless, military-neat office, I gathered that operability was the main challenge with the auto prize. "We are pioneers in this space," Cahill told me. "Trailblazers. And as a result we are having to do everything for the first time. It's hugely challenging from a resource standpoint."

Even after downsizing from a cross-country race to a single NASCAR track, the auto prize was still turning out to be a far more massive undertaking than the space prize. Back in 2004, when Burt Rutan's SpaceShipOne made its pioneering flight in the desert, there were no great corporate powers to alienate, and few political or regulatory thickets to hack through. The creation of an industry was the prize's whole point. In this case, the Foundation was meddling with one of the most entrenched and baroquely regulated industries on the planet. Also, the way the contest was structured, it demanded vastly more time, money, and manpower. With the space prize, the Foundation had just borrowed an airstrip from time to time, on a loose schedule that depended on the teams' readiness to fly. Now the Foundation was conducting an

Olympics. "We are *hosting* the competition," Cahill said. "You have to think of all the media rights and all the contractual obligations and legal complexities and layers and nuances." He gave me an example: The main sponsor, Progressive Insurance, happened to be a $15 billion corporation that wanted things a certain way. Cahill's team had created a style guide for the placement of corporate logos on the cars. It ran to thirty double-sided pages. The same guide for the space prize had been two pages. "Holy criminy," he said.

I got the sense that Cahill was doing the best he could in a tough spot, trying to manage powerful forces pressing down from above. But that wasn't his only difficulty, as became clear when he started talking about Oliver Kuttner.

Cahill brought up Oliver without me needing to ask. He described him as "exactly the little-bit-crazy entrepreneur genius guy" that the Foundation had set out to lure. It was in the organization's DNA to venerate a guy like Oliver: an inventor with an off-kilter idea. The dude was archetypal. At the same time, he was hard to deal with: loud, confident, voluminous across every axis. The inverse of Diamandis, who preferred to communicate in short bursts of type and talk and whose attention flitted easily and quickly from thing to thing, Oliver was forever sending epic, obsessive e-mails to Prize officials, offering his advice on how the Prize should be run. And man, the *size* of the guy. The sheer amount of clammy flesh that could shift in a blink if Oliver sensed a threat. It was a little unnerving how completely he'd lashed his dream to the Prize.

"That's what kind of scares me," Cahill said. "Like if he doesn't make it through Finals with one of those cars. Oh boy." He laughed. "Brace yourself for the maelstrom that is Oliver Kuttner."

14

Fit and Finish

hen Oliver was twenty-two, not long after he moved to Charlottesville, he made the biggest mistake of his life. He bought a building—an old roller rink on the edge of town. He wanted to start an auto-body shop, and he needed a place to put the cars.

He bought the roller rink for $5,000 down, then subdivided it into offices. The rent on the offices paid for his mortgage. The building rose in value. He refinanced, and the building wrote him a check.

Now Oliver decided to "get industrial" about real estate. He bought the former linen factory full of asbestos. He bought an abandoned coal tower. He bought several brick buildings on the town's Main Street, a promenade of struggling shops that was often deserted at night. He leased space to a pizza joint across the street from a little dive bar—the same place where a young

man named Dave Matthews used to play guitar for tips, before he
hooked up with a saxophonist and became the ninth richest lead
singer in the world.

In 1999, Oliver bought a beverage-distribution warehouse two
blocks south of downtown, on the wrong side of the old Chesa-
peake & Ohio train tracks. The plant was next to a public-housing
project. Few developers had ventured into this part of town before,
but Oliver saw promise. He got a bid to replace the walls of the
warehouse with massive panels of glass, but when it came back at
$300,000, he decided to design and fabricate the glass himself, and
got the job done for $60,000. The Glass Building, when it opened
on September 4, 2001, attracted new development to the neighborhood; ten years later, there's a pilates studio, a four-story building
of loft apartments, and an upscale restaurant. Oliver's buildings
on the promenade also appreciated in value; today the promenade
bustles at night with people drinking beer at outdoor cafés.

Oliver had a particular way of doing things. One day in 2001,
the chief city planner of Charlottesville, Jim Tolbert, walked over
to a construction site on First Street, right on the edge of the
promenade. Tolbert was shocked to see "this crazy German, with a
shovel, in a hole." At the time, Oliver was beginning construction
on his biggest development yet, the Terraces. Oliver had bought
the existing building on the site. He intended to renovate the
building for stores and offices, but before he could do that, he had
to excavate. He couldn't excavate normally, though, because there
was a Foot Locker store on the ground floor, and he hadn't been
able to get Foot Locker to give up their lease. So Oliver decided
to begin the excavation anyway, by digging underneath the Foot
Locker, bracing the hole with two-by-fours and concrete blocks.
There has to be a regulation against doing that, Jim Tolbert thought,
but he couldn't put his finger on what it was.

Tolbert had a lot of trouble getting Oliver to understand the
way the development process was supposed to work. You filed a

site plan with the city. You answered questions from the Board of Architectural Review. If you made a change to the building, you filed new paperwork. Oliver often resisted.

"I told him one time, he should have been building four hundred or five hundred years ago when they didn't require plans," Tolbert says. "He didn't want plans. He just wanted to do it."

Crucially, Oliver wasn't avoiding plans so he could cut corners. On the contrary, he was a perfectionist who approached development the way an artist would, lavishing time and money on fine details: a stained-glass window in the back of one of his buildings on the promenade; a series of elaborately crenellated, terraced roofs at the Terraces.

"I learned that I didn't have to worry about him building to code," Tolbert says. "He exceeded code. But he exceeded code at great expense in a lot of cases." Tolbert adds, "If we ever get a notice that the Bomb is coming, there are two buildings downtown that I would get in." One is the Terraces.

Oliver often burst into City Hall angry and flailing. He would belly up to the counter in the planning office and start cursing at Tolbert about red tape and regulation. The city of Charlottesville, he'd say, was holding him back, and if the bureaucrats didn't relax, he'd pull up stakes and take his business to Lynchburg. Then, half an hour later, Oliver would return to apologize. "Jim, you're right," he'd say. "I know you're just trying to do the right thing." Oliver could often seem like a caricature of the entrepreneur-as-asshole, the sort of wealthy, entitled goober who thinks the rules apply to other people but not to him. A giant toddler. But because he was such a creature of unbridled emotional experience, there was also a purity to Oliver that won over a lot of people who might otherwise have hated him.

A few years back, a woman named Courteney Stuart asked Oliver for a favor. Stuart was a reporter and editor at *The Hook*, Charlottesville's alternative weekly. She needed a male model for a fashion shoot. Stuart had gotten to know Oliver while reporting on big real-estate deals—"He's the only big-money developer in town who's always told me exactly what *he* paid, what *they* paid"—and figured that Oliver was such an extrovert that he'd probably do it. She was right. The plan was to do the shoot at a nearby historic estate. Oliver, who was slimmer back then, was going to wear outfits from the fanciest men's boutique in town. Instead of arriving at the shoot in one of the preselected outfits, though, he arrived covered in mud and concrete, fresh from one of his construction sites, dripping with sweat. He apologized for being late. "It looked like a rodent had given him a haircut," Stuart says.

The newspaper folks managed to clean him up. When the shoot was over, Stuart told Oliver that she had to go to the hospital to visit a friend and her newborn baby. Oliver said, "Can I come, too?" What Stuart remembers about Oliver at the hospital is how *happy* he was. He was dancing down the halls. The nurses wondered about this large, eager man pressing his face against the glass in the nursery. "Everybody's like, 'Who's your friend? He's dancing for your baby.'"

Oliver's desire to be liked made him different from his stepfather, Ludwig, who by now had risen to become CEO of the Hampshire Group, an apparel company that manufactured clothing for Dockers, Geoffrey Beene, and other brands. Ludwig also developed real estate in Charlottesville. Locals knew him as an important businessman, but a stern one, not to be messed with. There were Oliver people and there were Ludwig people, and they were not the same people. A close friend of Oliver, Jim Phelan, found himself on the same flight as Ludwig a few years back, and they got to talking about Oliver. Phelan recalls that Ludwig asked him if he thought Oliver would be successful. "I said, 'Yes, I do,'" Phelan says. "And his father said to me, well, he certainly hoped so, but he wasn't sure."

But Oliver's style of doing business was effective in its own way. He spent more time talking to filmmakers and carpenters and baristas than he did to bankers, building a following among the types of young creative people who might consider moving into the depressed neighborhoods Oliver wanted to revive; they started coffee shops and dress shops near his buildings, staffed his work crews, filmed his exploits with cars. Anywhere he planted the shaggy tree of himself, an ecosystem grew in the shade.

And because Oliver was so open to new people and new experiences, he tended to find value in assets others had written off. One year, he got a call telling him that some homeless people were squatting on the lawn of his coal tower. He drove over and found a group of twenty-somethings roasting a rabbit on a spit. They were train-hoppers. Instead of pressing charges, Oliver offered them jobs. He didn't know it at the time, but one of the men was a talented carpenter named Ben Jonas. Jonas had grown up on a farm in Utah, and he turned out to be a capable and fearless worker. He would go into some abandoned shell on Oliver's behalf, set up shop, clean out the piss in the corners, talk the crackheads into leaving, demolish the walls, renovate, and help Oliver flip the building. Jonas and his crew, especially a guy named Rocko, ended up doing so many jobs for Oliver that they joked that they should call the company THOR: Train Hoppers and Oliver Restoration. Oliver eventually made them partners on one of his projects, one-third each. Today Jonas owns more than eight buildings in Lynchburg and Rocko owns two.

Another Oliver hallmark was extreme candor. He would admit up front that he was out to enrich himself, which disarmed people and made them think that a guy so frank about his motives must be secretly altruistic. (Later, when asked about his motives for entering the X Prize, Oliver would say, "I admit, I got into it to win the money, and I am not proud of that.") The way he dressed, in jackets with holes and T-shirts stained with paint, made him seem friendly and vulnerable. He often collected rent from his buildings in person, and his tenants found it hilarious when Oliver showed up looking like he needed the money more than they did.

Elle Alston runs a clothing boutique across from the Edison2 workshop in Lynchburg, in a building owned by Oliver. She says, "If I went to him right now and said, 'Mr. Kuttner, I need twenty dollars,' he says, 'Miss Ellie, I don't have twenty dollars, I'm broke.' I say, 'This morning, Mr. Kuttner, I gave you rent. What did you do with it?' But he is every day broke." She laughs. "He promises me the world and gives me a spoonful of it. But he has a good heart."

THE POINT of Oliver's real-estate career was to indulge his passion for cars. The cars made the hustling worthwhile. A conservative estimate is that Oliver had more than one hundred cars stashed at his various properties at any given moment during the nineties. He owned a farm outside Charlottesville and kept cars there in horse stalls: Maseratis, Lamborghinis, Ghibli Spyders, Lotuses, DeLoreans, Ferraris, Bentleys, plus many models so rare that no one but classic-car freaks had ever heard of them: Bizzarrinis, Iso Grifos. He would eventually sell more than seven hundred of these exotic beasts. At one point Oliver cornered the global market on Bizzarrinis, powerful Italian sports cars manufactured by a former Ferrari engineer, Giotto Bizzarrini, who had left the mother ship in 1961 to start his own company. Only 139 Bizzarrinis had been made, and Oliver owned twenty-eight of them over the years, sometimes as many as six or seven at a time. People would call him from all over the world, complaining that his Bizzarrinis, at $90,000 each, were overpriced. "Yeah?" he'd say. "See if you can find one." (Today a Bizzarrini is worth between $400,000 and $700,000.) Oliver made so much money buying and selling classic cars that he was able to close his auto-body shop to the public and use it exclusively for his own cars.

Oliver also employed a team of mechanics who adapted sports cars into racing cars that Oliver drove himself. This was the purest and most childlike of his dreams: Oliver wanted to be a professional driver. He competed in some of the biggest races in the U.S., televised events like 12 Hours of Sebring and 24 Hours of Daytona. His racing team became known for doing a lot with a little. Oliver would spend $100,000 on a car and build it in two months, while his competitors would spend $1 million and take a year. Oliver enjoyed some success—twice he finished in third place in a moderately prestigious race called the Six Hours of Watkins Glen, in upstate New York—but his bids at racing glory taught him the size of the gap between him and top-echelon drivers.

The problem was partly physical. A typical race driver is lean. Oliver was a thick six-four. In every race, he was both the tallest and the heaviest driver, and the more he ate, the harder his teams had to work to compensate for his bulk. Eventually, after years of trencherman heroics—he would think nothing of ordering two full steak dinners at a restaurant and cleaning both plates—Oliver could no longer wriggle into the small, sweaty, clammy cockpits of his cars. "I didn't want to see the cars slowed down by a forty-eight-year-old fat guy," Oliver remembers.

His struggles with weight led to an insight: Maybe he wasn't the only glutton in the situation. For a long time, Oliver had sensed that cars were too heavy, especially American cars. He refused on principle to drive SUVs; if he needed to haul something large, like a couch or a desk, he'd throw it on top of his BMW 7 Series. For all his wealth and comfort, there was a side of Oliver that had always openly disdained luxury; it went back to his youth in Munich, when his father used to tell him stories about Germany having to build itself up from nothing after the devastation of the war. He felt that Americans were wasteful. Much of the thrill he got from selling cars and rehabbing real estate had to do with thrift, with giving new life to old and beautiful objects and ideas.

(Later, he would describe Edison2 as an effort to "take an obsolete technology"—the heavy automobile—"and reconstitute it.")

In 2002, he started a company, Mass Management, to market devices that would shave weight off cars. He had an idea for a brake system that would do the same job as a normal brake-and-wheel assembly but weigh 30 percent less. Oliver had come up with ideas for inventions before. Once, he had envisioned a modular bathroom that could be dropped into a building from a crane—the fixtures, the sink, the toilet already built in. But Oliver's inventions had never come to anything. The prototypes were sitting in his warehouses, lost in the maze of his possessions.

Not long after Oliver launched Mass Management, he was scrambling through downtown Charlottesville, carrying a check for one of his lenders. He entered the man's office and saw an attractive blond woman at the front desk—Kim, the man's daughter. Oliver took off a huge dumpy leather coat and set it down, knocking over a delicate terrarium. As he tried clumsily to restore the terrarium, Kim said, flirtatiously, "Nice job, guy, knocking over the terrarium." Oliver flirted back, at which point Kim's father whispered in his ear, "She has a kid."

"I do not see a ring on her finger," Oliver said. "All due respect, sir, this looks like a woman who can handle herself." Kim laughed.

Kim had never met anyone like Oliver. He owned $250,000 cars, yet his underwear was full of holes. In his new house in Charlottesville (he had sold his horse farm), he had a huge bathtub made of red porcelain and relaxed by taking baths. He was quirky, he had a big heart, and he wanted a big family, like Kim did. She was a single mother at the time, about to graduate from nursing school. She had a three-year-old son she was fiercely protective of. "He fell in love with my son," Kim tells me. "He's an amazing father. I mean, amazing."

Kim and Oliver were engaged in 2003 and married in 2004. They started having children. Things were good for a while. One year, Oliver bought Kim a silver Porsche Cayenne for her birthday. She opened a business downtown, Petit Bebe, a high-end children's boutique. But even as Oliver and Kim deepened their roots in Charlottesville, Oliver dreamed of escape. He would drive his luxury bus, the Starlight Express, to New York, and walk around the city, thinking of how he could break into the New York real-estate market. When he and Kim vacationed to Miami, Oliver would tour old hotels he imagined he could renovate. He often felt trapped in Charlottesville. That's why he considered his old purchase of the roller rink to be the biggest mistake of his life: It had marooned him in "a town with no people," a town where he lived in the shadow of his parents, who were immediately recognizable when they walked down the street. Beatrix had dyed her hair the color of an amethyst geode, fine and lustrous, framing a delicate, high-boned face. Ludwig wore vintage Armani. They were just so eccentric, so successful. In Charlottesville, Oliver couldn't help being seen as a rich man's son. And Kim as a rich man's son's wife.

Their lives didn't change in a major way until 2007, when Oliver read about the Automotive X Prize on the Web and ran out to his white Chevy pickup in a heat of inspiration. He took a gas reading, then left the truck idling for ten hours. Then he took a second gas reading. The next day, he repeated the experiment, but this time he left the cabin lights on while the truck idled. He wanted to see if there was a difference in gas consumption. Was there? Oliver doesn't remember. He can't even say exactly what he was trying to figure out. At the time, though, his actions seemed like the first thrilling steps of a quest. "I was collecting data," he says. "Part of me was thinking that there was some way with some evil magic dust that you could take an existing car and tweak it into these efficiencies."

Eventually he settled on the path that would lead him to the shop in Lynchburg and the Very Light Car: a strategy of purity (lightness, aerodynamics) and force (four cars instead of one). It all meant a lot of long hours, a lot of risk.

As Oliver toiled in Lynchburg, he sensed that Kim, back in Charlottesville, was growing more worried about their financial future. The real-estate business was lagging. Less money than ever was coming in, and much of it was flowing out to the cars. (In an e-mail, Kim tells me, "It was not an easy time . . . but the stress on our family and marriage was not in my opinion related to the money. . . . I would say that it was the endless hours spent away from our family that placed such a burden on us as a unit.") But Oliver thought everything would be okay. By the spring of 2010, as the guys in the shop were getting the cars ready for the X Prize track events, he was starting to get feelers from people who mattered, e-mails from curious individuals within large companies— not just automakers but plastics companies, engineering firms, and European manufacturers of industrial vehicles, all wanting to know more about the Very Light Car. He made friends in government. Oliver looked into applying for a loan from the federal government's Advanced Technology Vehicles Manufacturing Loan Program, the one that had approved $465 million for Tesla Motors, but he didn't seem to have much chance; the program was mostly funding electric cars and hybrids. But Oliver convinced Virginia senator Mark Warner to visit the shop, and the senator seemed to walk away excited, counting the Virginia jobs that a successful Edison2 could create.

If investment didn't come from government, it would come from private industry. Oliver still dreamed of "getting married" to a big company. He knew it was possible because he'd read about such a wedding. The bride was a guy he'd never met but respected absolutely: Elon Musk, the multimillionaire leader of Tesla Motors and Space X, who often gazed out at the nation from TV studios and late-night talk-show couches, discussing electric cars, rockets, and his dream of walking on the surface of Mars.

"That guy is *driven*," Oliver said. He saw a lot of himself in Musk. They were both passionate foreigners who believed in designing cars to fill a need, not to satisfy the market or focus groups. It was the Steve Jobs model. "A lot of times," Jobs once famously said, "people don't know what they want until you show it to them." And so far, Musk was making it work—barely. In 2007 and 2008, Tesla came close to shutting its doors multiple times; at one point, Musk had kicked in so many personal dollars that he had to borrow money for rent. But then, in 2009, he found a groom: Daimler, the German giant. For about $50 million, Daimler bought a 10 percent stake in Tesla. The auto industry "has to reinvent itself," Daimler's chairman explained at the time; the company was investing to gain access to Tesla's technology. Beyond that, Daimler hoped that some of Tesla's entrepreneurial boldness would rub off. "We want to go down new paths," the chairman said, "and we believe the combination of such a young, very ingenious company and a very experienced, longtime success-ful company is a good one in order to find new approaches fast." (Toyota later invested $50 million in Tesla as well.)

The Daimler-Tesla marriage proved to Oliver that the suit-ors were out there and were scared about the future. By now, the Obama administration had cranked the regulatory vise; all auto-makers would have to raise their fleet averages to 54.5 MPG by 2025, or else they'd have to pay penalties. Yet the big automakers were still moving slowly. The first two plug-in electric cars, the Chevy Volt and the Nissan Leaf, wouldn't be introduced in Amer-ica until December 2010, three months after the conclusion of the X Prize. There was no way to tell whether people would buy them. (When evaluated by the EPA, the Volt and Leaf would fall just shy of 100 MPGe—99 MPGe for the Leaf and 93 for the Volt, far short of the 230 MPGe that GM had predicted.) A stake in Edison2 could give some automaker a valuable hedge against the electric car, Oliver believed. If the electric car was Plan A, the Very Light Car could be Plan B.

He could help them. If he could reach them.

———

HE CALLED me in early May, two weeks after the end of Shakedown. It was the first I'd heard from him since the event. He started with some news about Edison2's search for a corporate groom. The meeting with GM went "*really* well," Oliver said. "I'm telling you, they are on this like a dog. It's *really* good. I'm telling you, General Motors is no longer the same GM. There were two naysayers, and I think the rest of them walked out of that meeting with smiles on their face."

When I talked to Brad Jaeger, who had also attended the GM meeting, he described it a little differently. He said there were three main types of reactions. One group of GM engineers thought the Very Light Car "was the coolest thing ever," Brad said. "They were crawling around on the floor in their suits to get a better look."

Another group stood back, circled the car, and made skeptical comments: The lights were in the wrong place, the windshield was too severely curved, and the car would never pass federal safety standards. This contingent thought the Very Light Car was interesting, but they didn't think it was a "real car," meaning they couldn't see customers driving it every day.

The final group said things like "If we wanted to make a car like that, we could do it, no big deal." They wrote off the accomplishments of Edison2 as trivial. Said Brad, "It was Not Invented Here syndrome."

Edison2 will never win over this last group. But Brad told me the second group—the one that believes the Very Light Car is not a real car—is theoretically persuadable. He said he can understand their objection. There are hundreds of small details that signal a car's car-ness to a consumer, tiny but crucial design touches that distinguish a car from a mere box of steel and plastic—everything from the shape of the key to the curve of the door handle to the texture of the steering wheel. As Aptera Motors had discovered from its focus group, if drivers find any one of these hundreds of details wanting, they won't buy the car, no matter how efficient it is. The Very Light Car's ignition key looked like the key to a golf cart. The door handle, designed to be as flush as possible with the door, to eliminate drag, was difficult to use. The steering wheel was a miniature model more typical of a race car. In industry lingo, the car lacked "fit and finish," a sense of tight, polished completeness.

In this sense, it wasn't a car at all. It was a tool. The tool had a single purpose: to win an efficiency prize. But it didn't have to stay a tool forever, Brad told me. The next version could be built to address the criticisms of the GM engineers, keeping true to the philosophy of the current car while softening its sharp edges. If Edison2 could produce a more user-friendly vehicle, one that looks and feels like a real car, it could win over the naysayers.

Right now, though, Oliver didn't have the time, the money, or the freedom to develop the next version. What was necessary to win investment from GM or another large company was at odds with what was necessary to win the X Prize.

This was the problem. He was in the middle of the race of his life, stuck with a series of best-guess prototypes he wasn't allowed to modify. According to X Prize rules, at this point in the competition, the teams weren't allowed to make certain major changes to the car: fuel type, chassis alterations, changes to the engine and transmission, the location of major components, etc. If a team wasn't sure if it was allowed to change something, it had to ask the Prize and receive approval in writing. The basic design of each car was locked, to prevent cheating and to make sure the teams weren't able to substantially modify their cars at the track to perform better in one event or another, which would defeat the whole purpose of the Prize—to identify a car that can do everything. The cars had to drive as they were until the Prize was over.

As a racer, Oliver found this constraint to be particularly

unbearable. In racing, teams change their cars all the time, based on conditions at the track, feedback from test runs, and any weaknesses they perceive in their competitors. The essence of racing is modification. Oliver had never been in a race where he couldn't make changes to the car. And as he now made clear to me on the phone, he'd been haunted by a persistent doubt: *What if I lose by a single MPGe?*

If he gets 99 MPGe, he doesn't get the money. And then, catastrophe.

What had initially shaken his confidence, weirdly, was hearing about the Illuminati car dancing through the cones at the Moose Test during Shakedown. Not that Oliver believed that a handmade car assembled in a cornfield was a threat to the Very Light Car. But its impressive performance, contrasted with its ramshackle appearance, hacked a divot out of his complacency, and all sorts of latent fears began to push in.

The main fear was that he wouldn't hit all the targets. There were so many trade-offs. Improve performance in one area and you worsened it in another. The most difficult trade-off for Oliver was between fuel efficiency and emissions. The electric cars didn't have to worry so much about emissions because their cars lacked tailpipes; the Prize calculated their emissions using a formula that took into account the coal (or other fuels) that had to be burned to produce the electricity in their battery packs. But Oliver had to leap the emissions hurdle with cars running on customized motorcycle engines. If the Very Light Cars ran too dirty, they'd be eliminated, regardless of how efficient they were.

The emissions of an internal combustion engine depend on factors like engine temperature, exhaust temperature, and the ratio of air to fuel. Like every other gas car on the road, the Very Light Car incorporated sensors that measured these quantities and fed the measurements into a computer that adjusted the engine on the fly to minimize the chemicals coming out of the tailpipe. Reducing emissions became a tedious process of "rearranging the deck chairs," to use Oliver's somewhat ominous metaphor—of paying attention to what the engine was doing and then reprogramming the computer. "It's not crazy art," Oliver said, "it's just work." And often the work involved burning extra fuel. You burned fuel to convert some of the toxic gases into less toxic gases, and you burned fuel to warm up the catalytic converter, the filter that scrubs the exhaust before it leaves the tailpipe.

Overall, Edison2 would have to waste anywhere from 10 to 15 percent of the fuel in the car's tank just to prove to the Prize that the car could run clean. So the engineers and mechanics needed to make up for that lost efficiency. They knew a few tricks around the edges. One was a technique called exhaust gas recirculation, or EGR, that took some of the inert gases normally expelled from the car's tailpipe and pumped them back into the engine cylinders. Akin to the rebreathing apparatuses used by SCUBA divers who explore deep underwater caves, EGR is used in many modern cars to help with emissions, but Edison2 also wanted to use it to increase fuel economy. EGR can help an engine operate more efficiently at low load, and the Very Light Car operated at a very low load, because it was so light. (Its engine was capable of 40 horsepower, but the car needed only 5 or 10 horsepower to propel it forward.)

Another trick was the choice of fuel. Instead of using gasoline to power the Very Light Car, Edison2 used a biofuel called E85, a blend of 85 percent ethanol and 15 percent gasoline. In recent years, U.S. automakers have produced a handful of "flex-fuel" models that can burn either gasoline or E85, but they also complain that there aren't enough gas stations that sell E85 to make it worthwhile; less than 2 percent of U.S. gas stations have E85 pumps. (Although the teams' business plans were required to "show that the national fuel infrastructure will support the vehicles," the Prize hadn't taken the business plans seriously as a judging metric; they were evaluated on a simple pass/fail basis.) Prize officials had determined that a gallon of E85 contained 71 percent of the energy of a gallon of gasoline, which meant that for the Very Light Car to carry the same amount of energy as in gasoline, it needed a greater volume of E85. More liquid. This would allow Edison2 to spray more fuel into the combustion chamber, helping to cool the engine, because the temperature of the fuel would be less than the temperature of the engine. E85 also burns cleaner than gasoline, so it would help with emissions. (The reduced energy content of E85 also meant that, in the math of the competition, Edison2 would need to travel only 71 actual miles on a gallon of fuel to achieve 100 MPGe.)

Would these tricks actually work? Oliver didn't seem sure. In early June, several weeks before the next stage of the Prize, Knockout, was set to begin at the Speedway, he got an e-mail from a technician at Roush Laboratories, an engineering testing firm in Michigan. Roush has equipment similar to that used by the EPA to test autos for efficiency and emissions. At Oliver's request, Roush technicians had strapped the car to a series of dynamometers, or dynos, which are like treadmills for autos, capable of measuring horsepower and torque as well as emissions and fuel efficiency. According to the e-mail from Roush, the Very Light Car had performed admirably on the emissions part. It would pass the Prize's emissions hurdle. This alone was an accomplishment. But on fuel economy, the car was "farther than I thought" from 100 MPGe, the technician wrote. On the dyno, the car scored just 87.9 MPGe.

"We will do 100," Oliver told me. "But barely. We might do 104. 106. But we cannot do 120. It's just not going to happen. You know how people need to look at this? You look at a guy like me. Think of me as a 20-miles-per-gallon or a 25-miles-per-gallon guy. If I cut my fuel in half, I'd probably be healthier and thinner. At that point I'd probably be a 50-miles-per-gallon human. If I cut my fuel in half again, I'd probably die of starvation or have major malnutrition. Do you understand?"

He called me a lot as Knockout got closer, sometimes twice a day, often late at night when he was driving home to Charlottesville from the Lynchburg workshop. A few times he started by saying he wouldn't remember any of this by the time he got home, then monologued for a solid hour while his Jetta slid through the hills of Virginia. He talked about money, about the weaknesses of electric cars. He talked about his wife, how great a mother she was to their four kids, how he was letting her down by spending so much time on the cars and so little time at home. He said he was speaking to me so openly because he wanted other would-be entrepreneurs to understand the difficulty of innovation. Each time he called, he seemed a little more nervous, a little more critical of the people running the Prize. He said they'd framed it all wrong. Instead of being too low, the 100 MPGe target was turning out to be *too high*, Oliver said. If an 80-mile-per-gallon car emerged from the contest, a car with nearly double the efficiency of a Prius, that would be a huge achievement, but the Prize would still deem it a failure.

"Their opinion is that the car companies are a bunch of idiots and a few innovative people can do it," Oliver said. "My opinion is that no one has done it because it's really hard to do. They're in this dreamworld. 'Oh, it's all easy, and forty people are going to show how it's going to be done.' I don't think the X Prize guys today understand how many teams don't understand the physics of it. Like, they actually *believe* all of us. And I don't even believe myself." He laughed.

I wasn't the only one hearing from Oliver around this time. As Knockout approached, he started blitzing Prize officials with e-mails and phone calls. Convinced it would be a game of inches, he pleaded for leniency on a range of yet-to-be-decided issues. For instance, ballast was still a mystery. Edison2 hadn't been told how many pounds of ballast it would have to carry in each car, to make up for safety features that are found in normal cars but not the X Prize cars. What should the weight penalty be for not having an air bag? Oliver sent Brad to a junkyard. Brad weighed every air bag he found, along with other safety items. Oliver sent the list of weights to the X Prize, arguing that the weight penalty for all safety-related features should only be 63.7 pounds.

Oliver also argued about issues of timing: When should the range test be performed? Early in the morning, when it's cooler (Oliver's preference), or midday, when it's hotter? Little things like this, Oliver believed, could be the difference between having someone win the Prize and having no winner.

He was working the refs—or so it seemed. The ambiguity of the rules made it difficult to say for sure. If it wasn't clear where the lines were, how could anyone accuse Oliver of crossing them?

"Knockout is going to be a bloodbath," he told me. "There's going to be nothing left. They're setting us up. It's terrible. It's the rudest thing in the world."

15

Little Red Corvette

"What's up next?" Simon Hauger asks the student members of the West Philadelphia Hybrid X Team.

"Knockout."

"What's our goal for the Knockout?"

"Don't get knocked out."

It's the weekly 3 P.M. team meeting in the classroom in the Auto Academy's garage. Simon wears his black team jacket with the diagonal lightning bolt down the back. The kids snack on multigrain crackers, organic cookies, and potato chips made of peas.

"The Shakedown was a bit more intense than we had anticipated," Simon says. "And it was an eye-opening experience because this is not the Tour de Sol. The Tour de Sol was a fantastic competition, but this is a $10 million competition . . . and they found a lot of things on our cars that needed to be improved. Which is really good. We expected that."

He continues, "On Monday I was describing the competition in what way, Sekou?"

Sekou Kamara says, "There's like a two-tier—a Group A and a Group B."

"And where did I feel we fell?"

"Right in between."

"We definitely had a better design and were better prepared than the second-level teams," Simon says. "But standing next to a multimillion-dollar company's vehicle is a reminder of the seriousness of this competition. We had a hundred-page business plan to develop. And hundreds of thousands of dollars to raise. And our families to try to keep intact. So there are times where we're saying, *What are we doing to compete against Aptera?*"

The kids listen in silence, eating their crackers.

Simon is trying to lower their expectations without making them despair. He no longer thinks the team can win. But he does think they can get to the Finals. They have two cars. One can break or fall short on mileage or emissions, and as long as the other one hits the targets, the team will be invited back in July. And then, who knows?

"We've always beaten cars that were more highly engineered," Simon tells the kids. "It's like taking a Corvette and racing a Bugatti. Right? The Bugatti on paper's going to win. But if one of the sixteen cylinders or four turbos or seventeen other things that makes a Bugatti a Bugatti breaks, guess who wins the race? Corvette."

WEST PHILLY almost didn't make it to Shakedown at all. They almost didn't finish the cars. Over the winter of 2009 and 2010, more than six and a half feet of snow fell in Philadelphia. This was more snow than fell in Boston, Chicago, and Anchorage, and it disrupted the school calendar and therefore the team's schedule.

West Philly pushing the Focus.

On some snow days, when work needed to be done on the cars, Hauger operated an ad hoc bus service, picking up kids at their houses and driving them to the garage.

They had a lot to do. Their business plan, for one thing. They were counting on the business plan to bring them even with other teams. If they had an edge in the Prize, it was the practicality of their cars.

In a methodical, thirty-two-page document, West Philly laid out the structure of what they called the EVX Company. The company would produce not one hybrid but two, both aimed at young, style-conscious, urban consumers and priced significantly lower than comparable cars: $25,000 for the Focus, about the same as a Prius and $15,000 less than the projected price of a Chevy Volt, and $62,000 or so for the GT, about half the price of a Tesla Roadster. The cars' major components would come from U.S. companies like Ford and Harley-Davidson, emphasizing the EVX Company's "goal of defining the 'new' American-made car."

(The word "American" appeared twenty-three times in the plan.) To manufacture the cars, EVX would transform an old shipyard on the Delaware River into a 45,000-square-foot production facility that would eventually employ 129 Philadelphians. Kids from the Auto Academy would intern there; after graduation, they might work there, too. The idea was to create "green jobs"; one of the team's leaders, Azeem Hill, had studied the writings of Van Jones, then President Obama's green-jobs guru.

"It's an education-to-green-industry pipeline," Hill told me. "It's telling the auto industry, not only stop building dirty cars, but you have the money to build a school and have your workers made *for* you."

West Philly also mapped out ambitious goals for the cars themselves, including last-minute changes to make them more efficient and reliable. Right after the New Year, Simon began to replace the batteries in both cars. All along, the team had been using lithium iron phosphate batteries packed into milk crates; now they would use lithium-ion batteries of a different chemistry, made by a Pennsylvania company, packed into aluminum boxes. The new batteries had about 15 percent more energy per pound, and Simon made the battery pack bigger at the same time, so the Focus gained energy. But the switch was labor-intensive. It wasn't like swapping out the 12-volt battery in a normal car. The cars had to be effectively taken apart.

Not only that, the software that controlled the cars had to be rewritten. On the Focus, this job fell to Keith Sevcik, a thirty-year-old who had just earned his PhD in engineering at Drexel University, Simon's alma mater. Keith had worked in a robotics lab, testing autonomous vehicles. I asked him once if he had ever crashed anything, and he laughed softly and said, "Two robotic ATVs. And two robotic helicopters. And one or two drone planes. Not the big military ones. Small gliders." Keith had sandy reddish blond hair and a patient way of explaining things to the kids, who

The GT.

called him either Dr. Doogie, because of his youth, or the Master-mind, because he was able to control every aspect of the Focus with his laptop. He often typed with the laptop resting on the hood of the Focus.

The Focus was a complicated system, a patchwork of unusual components—a Harley-Davidson motorcycle engine, a magnetic clutch. Keith had to get all of the components talking to one another in a way that wouldn't cause friction or grinding or, worse, destruction. A driver would need to easily control both the Harley and the electric motor. And he would need a digital readout that would help him understand what was happening with the car.

After the new batteries and new code were installed in the Focus, the custom transmission suddenly broke in the GT. The team had to pull the transmission out of the car and ship it by freight to a special shop in Texas. The repairman told Simon that first gear looked like somebody had beat it with a hammer. (It turned out that the transmission had been missing a pilot bearing, a small donut of stainless steel.) Meanwhile, in order to put 100 miles on the odometer before Shakedown, as required by the rules, West Philly ran the GT in a hobbled configuration, with a backup transmission and no body. The adults took turns driving it in the cold, dressed in hoodies, hats, gloves, and long underwear.

As for the Focus, it logged the required 100 miles, but no more. There was no time to do any additional testing.

On the eve of Shakedown, in April, Simon tried to reassure his team. "Everything we learn about the car, especially from the people who are going to be judging it, is invaluable," he said. If the car stopped running, that was okay. It was information. He told Keith, "Let's push it till it breaks."

The GT, much to West Philly's surprise, ended up passing the Shakedown's braking and acceleration tests without a hitch. Not so with the Focus, the car they had thought was in better shape. With Simon at the wheel, the Focus failed its first two attempts at acceleration. Simon employed both the electric motor and the Harley to try to reach the required 60 miles per hour, but the car topped out at 58. The team worked on the problem all night in the garage; Keith tried to debug his program, comprising four thousand to five thousand lines of code. The next morning, Simon went back to the starting line and gunned it toward 60, and again he came up short. Maybe there was a bad cell in the battery pack, maybe an error in the computer code. After a fourth unsuccessful attempt, West Philly decided they had no choice but to freeze out the electric side altogether and use only the Harley. They had to power a sedan with a motorcycle engine. But the Harley refused to start.

After several more hours of confusion and struggle, the team finally got the electric side and the Harley working in sync, and Simon broke 60 miles per hour within the required 15 seconds, passing the test. But the glitches had rattled him. The Focus was temperamental. Not to mention heavy. It weighed 2,800 pounds.

The GT weighed 2,600. The Aptera 2e weighed 2,300. The Very Light Car weighed 800.

"Simon, our car is a tank," observed Big Mark. "You might as well put a turret on it."

NOW, IN Philly, as Knockout approaches and more reporters start to write about West Philly's X Prize bid, Simon starts to get some funny calls.

"Hey, Simon," one caller says. "It's Congressman Fattah."

"Hi, sir," Simon says, taken aback. Chaka Fattah is one of the most powerful politicians in the region.

"Wait," the congressman says, "I got some people here. Couple guys from NASA. Tell them about your program."

"Um, we build hybrid cars with kids in Philadelphia."

"We're gonna come to see you."

And they do. One afternoon in June, the chief technologist of NASA tours the garage with Congressman Fattah and several local TV news cameras in tow. The chief technologist speaks to the kids about the need to inspire young people to pursue careers in science. He also mentions a new propulsion system that NASA is trying to fund. Fattah is about to become the senior Democrat on the subcommittee that oversees NASA funding. The next day, there are pictures of the students standing with the chief technologist and the congressman on NASA's website.

The following week, Michael Nutter, the mayor of Philadelphia, visits the garage with his entourage. He climbs into the GT, listening intently as Sowande Gay leans into the window and explains the features.

"Does this have an iPod dock?" the mayor asks. Sowande shakes his head. "We're going to have to fix that," the mayor says, laughing.

On a weekday morning in June, the week before the start of

the Knockout phase, Simon gives the mayor a ride through Center City in the GT. They pull up in front of City Hall, where a crowd of students, parents, and TV reporters await, along with two Philadelphia Eagles cheerleaders, a blonde and a brunette, both dressed in black Lycra, waving pom-poms.

The mayor has set up this pep rally. He moves from the GT to a podium. The team, he says, represents the sort of positive story about the city that the media should be telling. "I fully believe they are going to win the competition," the mayor says. He beckons to Azeem Hill.

"We set out to prove that high-school students can compete on a national stage," Azeem says in a brief speech. "And we're doing it."

A few minutes later, the crowd disperses to the sounds of vintage Prince:

> *And you say what have I got to lose?*
> *And I say*
> *Little red Corvette*
> *Baby you're much too fast . . .*

16

This Whole New Scale

After Shakedown, the chief engineer of Aptera Motors called all the company's division heads into a room. He pulled out a stack of eleven pennies that he'd wrapped in lime-green tape. Here, he told them, pass this around. This weighs an ounce. There are 775 parts in this car. Take an ounce out of every part and you save 50 pounds. That alone would increase the car's range on a battery charge by 5 miles.

Then Tom Reichenbach lit into them. They'd forgotten the middle name of the car, he said. *Safety + Lightweight + Aerodynamics.* That was the formula in all of their marketing, repeated ad infinitum. "It's right there on the front of our website!" he said.

The car, he said, had gotten too heavy. It was supposed to be about 2,100 pounds. The bare-bones prototype known as PP8 was about 1,970 pounds. A more fleshed-out prototype was 2,050. But the latest prototype—the one Aptera had brought to Shakedown, and the one it intended to take into production—had come in at 2,300 pounds: a 200-pound disappointment.

The car had gotten heavy for the simple reason that Aptera had made it in a hurry, under extreme deadline pressure, and when you make a car in a hurry, you make choices based on what's easy instead of what's light. It's the same with food: If you eat only foods that are convenient, you get fat.

Luckily, few in the media had yet picked up on Aptera's weight problem. The bad coverage about the company had all been related to the car's disastrous Moose Test at Shakedown—the door popping open, the cones knocked over, the forty attempts it took to pass. Bloggers and commenters had been brutal:

> If I were one of the sorry saps that put money down for one of them I'd be beating down the door to get back what I could.

> Anyone who wanted to use the Aptera as a *safe* family car has just crossed them off their list.

> So the Aptera ejects passengers automatically to decrease weight thus increasing mpg efficiency?

Tom had dismissed the criticism. Improving the car's performance on the Moose Test would be a relatively simple matter of tuning the suspension—adjusting the springs and dampers, which he hadn't had time to do before the event. But weight was harder to remedy. Something had gone wrong, and now it needed to be fixed before Knockout. (Prize officials were allowing Aptera to make changes because they weren't modifying any major systems.) The future of the company depended on it. Their jobs depended on it. They had seven weeks.

"They understood," Tom recalls later. "What they didn't understand is that I was gonna drag 'em through holy hell."

EARLY IN the summer of the X Prize, a handful of Aptera's remaining employees, including Tom, relocate to Michigan. They buy a beat-up van and rent a cheap condo in Ann Arbor. Every day, they drive thirty miles from the condo to the place where they work on the car, a race shop called Pratt & Miller. Pratt & Miller is famous for building custom race cars for all the General Motors brands. It has recently begun to diversify into electric vehicles. As one of Aptera's sponsors, Pratt & Miller is lending the expertise of its employees and the use of its 100,000-square-foot headquarters.

On the morning of June 17, four days before the start of Knockout, I join the Michigan-based Aptera crew outside of Ann Arbor. "Hop in," says Tom, at the wheel of the van. He introduces me to three other occupants, all young: Brian Gallagher, a bearded electrical engineer; Richard Fabini, an engineer with a birdlike frame; and Luis Arias, the company's welder and fabricator. They're heading to Pratt & Miller to work on the car.

"These guys are really fun," Tom says, meaning the Pratt & Miller guys. "They know how to compete." He says that the Prize, so far, has been "baptism by fire," but "that's a good way to do it. It really accelerates your progress." The downside is that "your triumphs and tragedies are other people's entertainment."

Brian says, "Dude, if you've got someone hating on you, you're doing something right."

"That's right, dude," Luis says. He's wearing a T-shirt with a hot rod on it, and there's a colorful tattoo on his arm showing a skull flanked by burning pistons.

I ask if Pratt & Miller, famous for its racing victories with gasoline cars, knows anything about electrics.

"They're learning about EVs at a rapid pace," Tom says. "They're an integral part of the team, wouldn't you guys say?"

Brian says that one of the Pratt & Miller race engineers told him, "I can't wait until we win, guys."

"Yeah, he's got the bug," Tom says. "We've all got the bug."

Tom is half geek, half racer. In the early eighties, he earned a masters in mechanical engineering at the University of Kentucky, studying under a man who figured out how to heat pig barns in the winter using only solar power. "He could guarantee that he could keep a pig barn at 55 degrees all winter long," Tom says of his advisor. "He was actually doing something. I like to get things done. I built two collectors for him. Spent more time on the roof of the agricultural engineering building than I'd care to admit. My wife would ask me if I'd ever get off the roof." Tom also has a deep background in racing; in his twenties, he traveled around the country with his wife, towing his race car from track to track. He went on to work at Ford for twenty-six years, eventually leading part of the company's prestigious GT racing program. To hear Tom talk about it, the bug is strictly a racing thing: a deep compulsion to prove the car's worth on the track in the face of vast and oppressive skepticism.

When we arrive at Pratt & Miller, located in a nondescript office park, we have to watch where we step when we enter the lobby because the floor is lined with so many racing trophies. Luis, the fabricator, carries a backpack with a flame-decorated welding helmet looped through the straps. We soon emerge into a cavernous series of garages. We walk through one that houses the Corvette race trailer; on the wall is a picture of a platinum blonde in a pink bikini. Then we come to the familiar Aptera trailer that so dominated the Speedway skyline at Shakedown. It's been loaned to Aptera by Pratt & Miller. Next to it is the 2e prototype, as well as an earlier, dingier-looking prototype—the PP8.

"Welcome to command central," Tom tells me. "Just don't pull out any cameras. Forbidden in this place."

MONTHS BEFORE, I'd mentioned to the Edison2 mechanics that Aptera was working with Pratt & Miller. They reacted with visceral despair.

"Pratt & Miller?" Reg Schmeiss said. "If you're a wanker and you went to Pratt & Miller, you're pretty fucking good."

"We're fucked," Bobby Mouzayck said.

Now, as I tour the Pratt & Miller facility, I understand their reaction. The first floor is lined with offices in which young men with earbuds gaze at computers, manipulating 3-D models. Along one wall, casual as a candy machine, is a 3-D printer that makes mock-ups of parts by melting and molding some kind of resin. Out in the machine shop, a noisy industrial space with 50-foot ceilings and a floor covered with metal shavings, computerized milling machines and lathes carve away at billets. There's also an in-house, climate-controlled "carbon shop" where eight guys in black T-shirts and black jeans work on the underwing of a Corvette race car, cutting pieces from giant spools of carbon fiber and Kevlar and laying them by hand into molds. This place is everything that Edison2's shop is not: vast, lush, and full of the latest and greatest toys.

Back in the offices, I'm introduced to Julie Casler, a structural analyst. She says she's been helping Aptera with "component optimization"—making each part as light as possible without sacrificing structural integrity. On her computer, Julie pulls up an image that depicts four different 3-D renderings of the car's knuckle, a part that allows the car to steer. It looks like a robot claw. Each of the four pictures has a caption listing its weight:

BASELINE DESIGN . . . . . 5.62 kg
ITER 1 . . . . 3.78 kg
ITER 1b . . . . 3.83 kg
ITER 1e . . . . 3.95 kg

"We can simulate load using this software—cornering, braking, hitting a pothole," Julie says. "The program eats away at the design and makes it thicker and thinner, and the parts are built here." She says she has optimized forty-six different parts, helping Aptera shave about 200 pounds off the car altogether. What Edison2's engineers are just guessing at, Pratt & Miller has a way to simulate.

Back at the Aptera garage bay, the 2e is up on jacks. Luis picks up a screwdriver and focuses on a newly installed solar panel on top of the car. If Tom can use that solar power to light up the car's instrument panels, he won't have to pull from the battery pack. "It's free power," Tom says. "You don't want to let it go."

"This is like a Christmas present right now," Brian says. "Look at those cells. Those cells are sexy."

"Only an electrical man would say that," Luis says.

BEFORE THE guys all caught the bug, Tom had to explain to them what it meant to go racing. This was a challenge that Oliver Kuttner never faced. Everyone at Edison2 comes from racing. But Tom's employees are more eclectic. Brian used to work in biotech. Marques McCammon designed cars at Chrysler and Saleen. Richard built high-efficiency vehicles at the University of California–Berkeley. Luis was a California dragster kid. And then there is Paulus Geantil, who arrives today at Pratt & Miller in a separate car.

Paulus, an exuberant engineer with floppy brown hair, epitomizes the culture of Aptera. An accomplished computer programmer, he's the guy who wrote most of the code that controls the car's electric brain. In his free time, he makes robots, but not just any robots—robots that make other robots. He rents space for his robots in an airplane hangar, which is how he ended up coming to Aptera in the first place. One of the guys he shared the hangar with happened to work for the company, and they got to talking. Paulus thought, *Who cares about cars?* "Automotive already seemed so hashed out," he tells me. "My gosh, this is a culture of stagnation. And ultraconservative decision making." But then he started thinking about his father. A physicist, he had always encouraged his son to keep an eye out for what he called "earthly problems." It suddenly hit Paulus: *This is my earthly problem.* "If someone had told me I'd be working on a car, I'd be like, that's *crazy*," he says.

The other big difference between Edison2 and Aptera is time. Aptera is several years ahead of Edison2—and all the other X Prize teams—in figuring out how to get its car to market. In some ways, the Aptera employees see themselves as older and wiser versions of their Prize competitors, a fact that becomes clear when the team breaks for lunch and drives to an Applebee's not far from Pratt & Miller.

"I look at all these other teams and I think they're *exactly* where we were five years ago," says Brian in the restaurant. "We were so naïve. We thought, yeah, man. We can do this, and we can do this *better*. And we *did* it. And we said, okay, it *can't* be this easy to do this." Around the table, heads nod. "And that's how it started unraveling. This whole new scale opened up."

I ask Tom what he thinks of the Edison2 Very Light Car.

"Their cars are too spartan," he says. "No one would buy one."

AFTER LUNCH, Paulus, Richard, Luis, Brian, and Tom spend hours testing the 2e's electrical system. They want to make sure it's working properly before they take the car for a drive in the rear parking lot. While Brian and Richard probe electrical pins with a handheld voltmeter, measuring levels of current, I chat with Paulus. He says he's interested in "biomimetic solutions" that look to nature for inspiration. He points to the 2e on the side of the trailer. "This is Nature's penetrator," he says, referring to the car's phallic form. "This is the shape that's meant to *go through other things.*"

With the sun beginning to go down, the Apterans close up the car and push it to the edge of the garage. Tom spiders into the cockpit and takes off across the parking lot. The engineers swivel their heads to follow the car.

"Hey, you know?" Paulus says. "I like what I'm seeing."

Brian says, "Look at this, dude. I don't see jitter."

Tom comes back toward the group, then stops and turns and goes the other way. Several bursts of speed later, he pulls over and gets out. It feels smooth, he says. "Just the slightest t-t-t-t of jitter." He asks Brian if he wants to take a test drive, adding, "Don't break it. I gotta say that."

As Brian accelerates to 35 or 40 miles per hour, Tom asks Luis what he thinks.

"Pretty cool," Luis says. "Looks like a spaceship."

"It doesn't feel a lot faster," Tom says. "It's pretty smooth though."

"Regen works, too," Brian says of the regenerative braking system, returning from his test drive. "Regen feels good." He gives up his seat to Paulus. A few moments later, Paulus comes back grinning.

"*Awesome*," Paulus says. "You can use the whole pedal. That makes the whole difference. And 100 percent regen feels good."

Another Aptera engineer hops in and guns it. The tires squeal. Tom smiles.

"He's gotta try and break it," he says. "Engineers—you give 'em a toy, they gotta try to break it."

17

Lungs Cannot Oxidize

The evening sky above Kevin and Jen's house turns the color of rust. Kevin walks to the edge of his driveway. Beyond the far tree line, he sees the wind turbine that supplies a small percentage of his electricity. The local electrical co-op recently built it to generate clean energy for its 5,800 members. The turbine, made in Europe, sits atop a pile of coal residue at an abandoned mine. Its blades are spinning so rapidly that they start to blur. A finger of cloud dips from a thunderhead, touches one of the trees, and recoils as if the tree is scalding hot.

When the storm comes, it dumps two inches of rain on the corn and snaps a gnarly limb from Kevin and Jen's river birch, depositing it on the tarp over their pool. After the rain, the sky turns a luminous orange pink, and the fireflies blink in the corn like error codes.

With the thunder growing fainter, Team Illuminati work into the night rewiring the car's electrical system, which is plagued with problems. The car hasn't been on a test drive in a month. The turn signals, the struts, and the inverter are all on the fritz. The inverter especially. An inverter is a sensitive gadget found in electric cars that switches one kind of current, the kind produced by the batteries, to another kind of current, the kind required by the motor. It does this at an incredibly high speed. For the inverter to run at peak performance, it needs to be "auto tuned" to the motor, so that both components run in perfect sync, but the inverter keeps on failing to complete an auto-tune cycle, which means the motor isn't working, either. Kevin was on the phone earlier with the guy in Oregon who sold him the inverter. He told Kevin that if it wasn't working, it wasn't the inverter's fault, it was the fault of whoever had wired the system.

Now the team is almost out of time. The Knockout phase won't begin for another eight days, but because Illuminati failed certain tests during Shakedown—0-to-60 acceleration, braking, durability, and others—the organizers are requiring them to arrive at the Speedway five days early, to perform those tests again. Only then will they be allowed to proceed to Knockout. (A number of other teams have also been asked to come back early for similar reasons, including a few European teams that missed Shakedown because of flight cancellations caused by the eruption of Iceland's Eyjafjallajökull volcano.) The upshot: In only two days, Illuminati's presence is required at the Michigan International Speedway. The electrical system has to be working flawlessly, because Nate, the electrician, hasn't been able to get time off his job to attend Knockout. He won't be in Michigan for the first week. Kevin will be on his own. This might be a problem, because Kevin doesn't understand the wiring like Nate does. "Nate's a really bright guy, lucky for me," Kevin says. "He's doing a much better job than I could. Look at those wires. BLAAAAAAAA."

It's hugely risky to rebuild major systems of a prototype automobile at the last minute like this. You can't control the element of uncertainty. As soon as one system is fixed, another breaks. "It's a cascade effect," Kevin explained once. "Okay, now you've let the lion out of the cage. So you bring in the lion tamer with the whip. And if that doesn't work you bring in the guy with the tranquilizers, and if that doesn't work, you bring in the guy with the shotgun. And if *that* doesn't work, you bring in the guy with the Mack truck and tell him to swerve a little to the left."

Jen watches Kevin work. The light is sawdust-colored. A message on the whiteboard reads "Innovation does not come from a good night's sleep."

The car isn't running, she thinks. *It isn't good enough. This is it.*

Then, *Kevin thinks it's good enough. Maybe he's blowing smoke at me.*

She goes to bed. Everyone except Kevin and Nate goes home. Kevin starts coughing. He says he feels feverish.

At 2 A.M., he leaves the shop, walks the twenty yards to the house, and crawls into bed. Jen is awake, listening to the thunder.

THE NEXT day, June 14, Jen can't concentrate at work. She can't stop running through a list of all the ways the team is screwed.

If the car isn't fixed by tonight, they're out of the competition.

Even if the car *is* fixed, it might not stay fixed, because bugs always crop up at the track, and Nate won't be at the track next week to squash them.

Nate is the one who knows how to talk to the car. Not Kevin.

Kevin's sick. His body has succumbed to the stress.

I can't watch, Jen thinks.

When she gets home, she changes into a T-shirt. THE APATHY COALITION, it says. JOIN US. OR DON'T. WHATEVER. She walks to the shop. The team members are all there, making a final push to get the car working. They're also sugar-loading on junk food: strawberry zebra cakes, cosmic brownies, raspberry chocolate chunk gourmet muffins, Twinkies, donuts.

Nick, Kevin's father, pulls a piece of painter's tape from the car. A scab of silver paint comes off with it. Improvising, Kevin tells Jen that if she grabs a silver Sharpie and scribbles on the spot where the paint peeled off, it will look good as new. She obliges, carefully drawing with the marker, then steps back and regards her work. She makes a sour face.

"You're absolutely right," she says. "Silver Sharpie looks good as new. It works really well."

"Don't cry, Jen," Nate says, teasing her.

Nate went to town earlier to buy replacement fuses. Jen asks him how much Kevin owes for the fuses. He tells her $54.

Jen digs into her pocket and shoves a few bills into his hand.

"That's fine," he says.

"No, it's not." Jen runs to the house to get more money.

The electrical system still isn't debugged, and the liquid-cooling system that keeps the motor and inverter from overheating is in tatters, but Kevin wants to go for a test drive before it gets dark, so with Nate's help, he pushes the car out of the shop and onto the gravel driveway. Nate takes the shotgun seat. Jen climbs into a backseat through an open gullwing door and buckles her seat belt. I sit next to her.

"You got your license on you, Kevin?" Jen says.

"No, I do not."

"No license, no insurance, no registration?" she jokes.

Jen runs her fingers along a rectangular bulge next to her seat. It's one of the battery boxes, covered with speaker felt, as are many other interior surfaces. Kevin bought speaker felt instead of carpet or leather because it was cheaper. Sitting in the car feels like sitting inside a giant guitar amp.

Kevin turns the ignition key and puts his foot on the gas. The car goes shooting down the county-line road. We smell something plasticky and coppery. The regen brakes kick in, and the car slows to a stop. We all open our doors.

"Oh, shit," Jen says, as an animal howls in the distance. "I don't know what that is."

"We have teeth, too," Kevin says, smiling. "What are you complaining about?"

Crickets chirp, cicadas buzz.

"You know," Kevin says, "I bet we could drive it at a comfortable speed with the doors open."

The road starts to slide out from beneath us.

Wings extended, doors open, the car is moving now—10, 15 miles per hour—and it feels less like we're being pushed (or in this case pulled) and more like we're being rolled down a hill. No internal combustion engine, no black smoke, only a little electromagnet spinning faster and faster—40 miles per hour, 45. A feeling of impossible luxury. The ultimate convertible ride. At 50 miles per hour I can't hear anything except the whoosh of air in my ears as the corn blurs into a sheet of yellow. I've got this weird giddy feeling in my chest. *You can make an electric car by hand. You can make it out of steel and fiberglass and epoxy and plywood and fire.*

And then suddenly and with no warning Kevin pulls hard on the steering wheel, slamming Jen and me into our shoulder harnesses and whipping the car around 180 degrees.

"Accelerating still isn't as quick as it should be," Kevin says to Nate.

"Only got to 54 miles per hour," Nate says. "That sucks."

"What's that smell?" Jen says. "Holy shit. Probably brakes?"

"Naw, that was batteries," Nate says.

Kevin needs more eyes on the problem. He drives us back to the barn. He glances at Jen, and Jen nods and looks at me: time to get out. We pop our doors. Josh and another guy take our seats. Jen and I stand in the driveway watching the car recede until its two taillights blur into one, glowing in the blackness like a city viewed from space.

When the car returns, half an hour later, the guys push it

into the barn. Kevin opens the hood and puts his hand on the hoses leading to the inverter. They're warm to the touch. He chugs a sweaty Fresca and takes his ball cap off. "Look at all that dust and dirt that collects on the plastic," he says, grinning with ironic pride, as if he designed the car not to win $5 million but to gather lint from country roads.

"We gotta quit running it without cooling it," Nate says.

"Ya think?" Kevin sneers.

This is strange. Kevin doesn't get angry, ever. Nate looks the car up and down with blank, proprietary eyes, like a farmer taking the measure of his heifer.

"I hope we can get that accelerating," he says. "It seemed like it was slowing down when it was accelerating."

Kevin shakes his head. "Sumpin' ain't right," he says, going full redneck for effect. Everyone starts talking at once, offering suggestions, asking questions. Kevin waves them off.

"It's doing lots of different things, okay?" he says curtly. "When I try to shift . . . it made bad noises. The clutch pedal had a bad feel. We didn't have power for a second. We had to stop . . . and then everything was happy, so I didn't press the clutch pedal anymore. I tell ya, man, I was afraid to try it again. You might only get one shot."

Kevin is basically saying: This is a car that stops but doesn't go. When it goes, it doesn't go fast enough, and when you try to shift gears, which you need to do to accelerate, the clutch slips, and some kind of violence happens down in the guts.

There are now eight hours left to get the car working and loaded onto the trailer for the trip to Michigan. Eight hours to save two years of work and a $100,000 investment.

AT MIDNIGHT, with the men circling the car, I go into the house with Jen. We sit at the kitchen table in front of the tank containing

Jen's managuense cichlid, a rotund fish with leopard spots—"a real mean bitch," she says jokingly. She gets herself a beer from the fridge and hands one to me. Down the hall past the living room, I can see framed photographs of character actors on the walls—head shots of people like Bruce Campbell from *Army of Darkness* and William Shatner, collected at science-fiction conventions. A shrine to the tenacity of the underdog.

Jen is tired. The project has engulfed her life. Since 2009, it's basically amounted to a part-time job. Most of her X Prize hours have been spent on the computer, handling the team's website, social media, and press releases, designing the team's decals and T-shirts, and jumping through the flaming hoops of the Prize's paperwork requirements. But Jen has also helped with the dirty work: the messy, repetitive, smelly jobs that involved inhaling volatile chemicals in an enclosed space. She resents when reporters or friends assume that her role is to cook for the team. Along with other friends and family, she helped cover the car's layer of fire-retardant foam with fiberglass and epoxy, then smoothed out the rough spots with a body filler called Bondo. "Current strata of my lungs," Jen Tweeted during this phase: "Bondo dust, epoxy primer, second hand Pall Mall, & a dusting of black matte Rustoleum. My lungs cannot oxidize."

Now she sips her beer and starts talking about what's going wrong in the shop. There are some personal issues, she says. Thomas is too slow and doesn't work well with others. "Thomas rebuilt my Gremlin's engine. It took him nine months. When he got done, every bolt was painted either black or red. It was beautiful, but he could have *gestated* it in that time." She considers Nate. "Nate, he's never had to design the electrical system for a car and make it work in a competition, in front of the media."

And her husband? He's too optimistic. He's not like her. Kevin really believes all this stuff about the value of mavericks and the power of crazy ideas and the Prize as a level playing field.

He believes that nothing is impossible when you're building a car because all is engineering, and engineering, in the words of historian Henry Petroski, is merely "the rearrangement of what is." But what happens if the Prize fails to live up to Kevin's idea of it? What happens if the rules aren't fair—if it's not really the perfectly democratic and level playing field it purports to be? She fears he'll come home damaged, permanently diminished, and she'll be the one who has to pick up the pieces.

I ask Jen if she's traveling to Michigan with the team. She smiles without showing her teeth. "I can't do that again," she says. "I had a meltdown before. Three days of inspections? Jesus. I only went because Kevin asked me. I don't like competitions. I get stressed. My impending sense of doom." (She'll later change her mind and go to Michigan after all; Kevin wants her to be there, and she also feels a duty to live-Tweet the event for supporters back home.)

The mantel clock ticks, the fish tank burbles.

When we finish our beers, we walk back out to the shop. As soon as we open the door, we know something is wrong.

There are two black rectangles on the floor. The covers of the battery boxes. The covers are not supposed to come off, ever, for safety reasons.

We peek into the car. We see ninety-six lithium cells sheathed in blue, flashing yellow error lights at their terminals, red and yellow wires strung between them like clotheslines in tenements. It feels indecent to see them there, exposed, like the living, pulsing lobe of a patient's brain after a neurosurgeon has removed part of the skull.

"We lost power," Kevin says flatly. "We have five volts to run the car on." Five volts won't run an electric toothbrush.

While we were in the house, he explains, the electrical system suddenly lost power. When he and Nate cracked open the battery box to see what was wrong, the inside was filled with oozy, goopy stuff. Glue. Nate had used a hot-glue gun to cover potentially exposed conductors in the battery box, in compliance with X Prize safety rules. (Every terminal and connector had to be separately covered, so that even if you dropped a wrench in your battery box while it was open, you wouldn't get a short or flashes of electricity arcing through the air.) The melted glue meant that the batteries must have overheated during the test ride earlier that evening.

But why did they overheat? When Kevin and Nate sopped up the glue as best they could, they saw that, beneath the sticky stuff, there was a little melted steel bolt where a piece of copper should have been, in a part of the battery system called the fuse holder. Steel is a poor conductor of electricity. The steel bolt had broken the circuit. Instead of the electrons zipping happily through the copper, they were stopped by the bolt, generating heat, which melted the glue.

It's now too late to rewire the electrical system. It has to be patched, and quickly. To repair the broken circuit, Kevin needs to find some scrap copper lying around the shop and weld a piece of steel into the fuse holder that will hold the copper in place. If he can't do this, the car won't start. Not now, not ever.

Kevin knows it's a bad idea to attempt this repair. He doesn't have time to disconnect the batteries and make it safe for him to stick his hands in there. So he will have to work with a live, hot system—hot in the sense that a household electrical outlet is hot, in that it's ready to provide power the second that a circuit is completed. The fuse holder is jammed into a place where fingers won't easily reach, and if any part of Kevin's finger touches anything that's not the fuse holder, the broken circuit will be completed by Kevin's body. A potentially fatal shock will flow through him. He'll have to make the repair quickly and he'll have to do it on no sleep. Kevin understands the danger, but Jen doesn't, and he doesn't explain it to her.

"Not to be a killjoy, guys, but we got four and a half hours to get everything working," Jen says. She leaves the shop, walks back into the house, and goes to bed.

Kevin wheels the MIG welder over to the edge of the car. He picks up a length of scrap copper pipe from a household plumbing system. He sits on the floor next to the shotgun seat, left arm balled in a fist behind his back to keep himself from sticking his spare arm into the car in a moment of carelessness. Deciding that two years of work is worth possibly dying for, he pokes the fingers of his right hand into the electrified box.

Nate has just thrust his hand into something in the hood. He clutches a wormy tangle of wires. He moves his hand slightly and the whole car jiggles.

"Kevin's got his hands in there," Thomas tells Nate. "He'd just as soon you not be shaking the car, man."

"Yeah, yeah, yeah," Kevin says. "I'd like to keep my hands. My right hand, I'm no good without it."

The welder flicks on. Kevin jabs his finger into the space where he can't see. There's a noise, a puff of acrid smoke, the smell of copper oxidizing. Kevin pulls his finger back, walks around to the open hood, places both hands flat on the housing of the electric motor, stiffens, jerks violently backward, and yells "AAAAAAAAA," like he's been jolted. Nate cries out. Kevin stands up, chuckling.

"Not funny, man," Nate says, exhaling. "Not funny."

Kevin goes back to work. Sometime after 3 A.M. he starts quoting from the lesser, odd-numbered *Star Trek* movies. Thomas falls asleep in a chair.

Jen enters the shop at five, wearing sweatpants and sandals. Thomas is snoring. Kevin is on the floor holding a ratchet.

"Oh, hi, honey," he says.

Thomas abruptly snorts himself awake, loses his balance, and almost falls out of his chair.

"Time to bite the bullet," Kevin says. He picks up a voltage meter. This tool will tell him if he has placed the copper correctly. If he still has a car worth taking to Michigan.

"With any luck," he says, "we should have 307 volts."

He presses the tip of the meter to a battery terminal.

"THREE EIGHTEEN, BABY!"

Kevin opens a can of diet soda and then opens the garage door. The sun is coming up. The sky is a sheet of violet light, and grackles shake the branches of the pines. If Kevin buttons up the car in the next three hours and drives 70 all the way to Michigan, he might make it in time for check-in.

18

Voodoo Steve

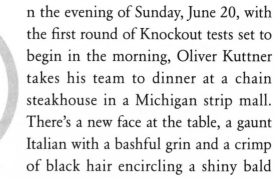n the evening of Sunday, June 20, with the first round of Knockout tests set to begin in the morning, Oliver Kuttner takes his team to dinner at a chain steakhouse in a Michigan strip mall. There's a new face at the table, a gaunt Italian with a bashful grin and a crimp of black hair encircling a shiny bald pate: Emanuele Pirro, one of the best sports-car drivers in the world. Pirro is a forty-eight-year-old veteran of Formula 1 and the five-time winner of 24 Hours of Le Mans. Ron has known him for years. Emanuele will be driving for the team from now on, in addition to Brad Jaeger. The gap in experience between Emanuele and the other teams' drivers, almost all amateurs, will be hilariously large. It's akin to hiring Tiger Woods to fill out your Sunday scramble.

At the end of the meal, Oliver rises from his seat. Fifteen Edison2 team members look up at the boss. All conversation stops.

"It's ours to lose," Oliver says.

His voice overpowers the radio in the steakhouse, which is

playing "Love the One You're With." His tone is icy and disdainful, like that of a presidential surrogate flicking away a question that's beneath the dignity of the office. Oliver says that except for the Swiss-made covered motorcycle, the E-Tracer, all the other top competitors are running into mechanical problems. "You're going to see a car self-implode, there's no doubt in my mind," he says. "The cars I thought had a chance really were much heavier than they were planning. I mean, they blew their weight by 50, 60 percent. There's nothing there that can do it on paper."

Oliver looks at P.K., his head mechanic, then at Ron, his head engineer. A fiery German addressing two meticulous Brits. Oliver stresses that for Edison2 to win, P.K. and Ron must commit to "the endless pursuit of not leaving anything on the table." P.K. smiles and squirms in his seat, either amused or slightly offended that anyone would need to remind *him* to pay attention to detail: P.K., whose Snap-on toolbox is better organized than a cabinet of butterfly specimens at a natural history museum.

Then Oliver jabs a meaty finger at Ron. "I would enjoy to shoot you on the moon for the most expensive seat-track system known to *man*." Not so long ago, Ron spent $40,000 of Edison2's money on ultra-lightweight seat rails that ended up saving a few pounds per vehicle. The memory of the seat-track expense—so recent, so fresh—is painful enough to garble Oliver's usually sturdy grammar: *I would enjoy to shoot you on the moon.* Ron laughs. Everybody starts talking and laughing at once about the seat tracks. Oliver seems to falter for a second, his weight sagging, but then he straightens.

"But those three pounds just might be the difference, so let's go."

The team stands in unison and files out of the restaurant, joking and horsing around. Oliver stays behind. He frowns, mumbling to himself, fishing in his jeans pocket for an un–maxed out credit card to pay for the fifteen steaks.

———

THE FOLLOWING morning, Monday, June 21, a white fog settles over the Speedway. Journalists and documentary film crews roam the garages. No one is welding or power-sanding. There's a lot more room to move than there was at Shakedown. You can drink coffee without fear that a mist of metal is settling into your cup.

Only twenty-one teams and twenty-six cars remain, including just six teams in the all-important mainstream class, worth $5 million. One recently eliminated team is Cornell University, which withdrew after discovering damaged circuit boards that signaled a problem with their electrical system. "We don't really want to take a chance with people's lives," the Cornell team leader told an X Prize film crew. "You know, there's a lot of energy stored in those batteries. It's the equivalent of many sticks of dynamite." Cornell's elimination meant that West Philly had now beaten an Ivy League school.

In a middle bay, more than a dozen West Philly students and seven adults in matching black polo shirts scramble to prep the Focus and the GT. Some kids press sponsor decals onto the paint; Azeem Hill pores over crumpled sheets of torque-bolt specifications. In the trunk of the Focus, Jacques Wells grips a warm soldering gun and looks at Simon Hauger, who thinks the fuel relay pump can be improved. He wants Jacques to solder the new relay.

"You're gonna strip this wire, right?" Simon asks Jacques.

Jacques cocks his head quizzically: *You sure, Haug?*

"If the car doesn't work, I assume full responsibility," Simon says.

I take a picture of Jacques working in the trunk. My flash goes off, and Jacques leaps back, startled. He puts his hand on his heart and exhales fiercely. "I see a flash around electric stuff, I get scared," he says.

A little later in the day, Jacques takes a walk through the garages, checking out the competition. The E-Tracer, he thinks, is pretty slick. "I *want* one." But Jacques doesn't see anything he likes better than the West Philly cars. The Very Light Car, the Illuminati car, the Aptera 2e—too futuristic, too impractical. "With the Harley, along with the electric motor, you get a hundred miles per gallon, easily," Jacques says, explaining the virtues of the Focus. "All the safety features are already there. All of the accessories are already there. You don't have to worry about any hard-wiring to make the wipers work. It's already *done*."

TOM REICHENBACH of Aptera stands in one of the garages, motionless amid the flux. He looks around and smiles. He moves his hand successively higher, from one level to another. "The level of competition is much higher than when we were here the first week," he tells me. "The competition is working perfectly."

Tom walks out the rear of the garage toward the Aptera tent and climbs into the Aptera race trailer. Munching on a handful of trail mix, he explains that the day after I visited him and his team at Pratt & Miller, Aptera took its slimmed-down car to the Chrysler test track. "We expected improvement," Tom says, glancing off to the side. "We got *de*-improvement."

After all that sophisticated computer modeling and optimization, all that expensive retooling of parts, after a summer of passing around the stack of pennies, the car's numbers actually went *down*. Tom sounds less upset than perplexed. How could the car feel better in every way and yet perform worse? "That's what's been driving us crazy," he says.

THE NEXT morning, at a closed-door meeting involving the Prize officials and representatives of the twenty-one teams, the officials review the rules of Knockout. It's basically a rerun of Shakedown, with a few critical changes. The cars will perform all of the same tests they did at Shakedown—0-to-60 acceleration, braking, the

Moose Test, plus a few extra ones—such as a cornering test on a skidpad—and a careful measurement of emissions gases.

The big difference is that the cars now have to perform three "driving cycles" on the oval. The cycles measure fuel economy, MPGe, and there are three so that the Prize can get an accurate idea of how the cars might drive in all sorts of conditions. Officials will place signs at various spots around the oval telling drivers when they should stop and what speed they need to maintain, and the signs will be different for each of the three tests. There's a 16-mile "city" cycle with lots of sudden stops and starts, a 90-mile "highway" cycle that's mostly cruising, and a 30-mile "urban" cycle that's somewhere in between. Each car will register a certain MPGe on each cycle, and the three cycles will be combined to get the car's final MPGe.

For the first time, there's also a range test, in which alternative-class cars will drive around the oval for 100 miles straight, and mainstream-class cars will drive for 200 miles. If they're experiencing problems, the cars are allowed to "pit"—to pull off to the side, or into Pit Lane, where mechanics can work on them—but some events require laps to be completed within a certain time, and teams that pit too often will incur penalties.

The tests will be staggered, with different cars doing different tests on any given day. During the runs that measure fuel economy, five or six cars will compete next to one another, at the same time. Because there are so many tests, and because they need to be judged so carefully, Knockout will take more than a week. But many of the teams won't make it that far. Unlike Shakedown, Knockout is an elimination round, so if you fail, instead of just learning about your vehicle, you go home. The targets are two-thirds of what they'll be in the Finals. For fuel economy, the cut-off is 67 MPGe. The range cutoff is 134 miles for mainstream cars and 68 miles for alternative cars. All performance statistics will be released on the X Prize website for the world to see.

What's more, if you fail a test or miss a performance target, you're eliminated immediately. You could be knocked out on the first day. The idea is to keep the contest moving—the point, after all, is to identify one and only one winner in each category. If you think you've been knocked out unfairly or in error, you can appeal the decision to a panel of three independent judges not employed by the Prize, but you have to accept the panel's decision as final.

The meeting of the officials and the team leaders ends with Oliver standing to deliver an impromptu speech. Minutes later, Simon Hauger recounts Oliver's speech for the benefit of several West Philly adults who hadn't been in the room. "You're all leaders," Hauger says, quoting Oliver, "so even though there are problems, you have to put on a brave face."

Simon rolls his eyes. "It's like, man, Oliver, it's just the second day," he says. "I was trying to figure out some smart-ass thing to say."

"Well, Oliver's the one who knows the correct way to get to 100," says Ann Cohen.

"Yeah," says another West Philly adult. "You rebuild the motor every 20 miles!"

THE SPEEDWAY is shaped like a D. The straight part of the D is the straightaway on the far side of the track, opposite the garages. The start/finish line falls in the middle of the curved part of the D. Overlooking the D are grandstands that seat twelve thousand.

This is where most of the action happens at Knockout. At the start of each economy run, the teams push their cars from the garages toward a long, thin, two-story building parallel to the grandstands. The building contains offices where Speedway employees monitor the track. Beneath the building is a tunnel, above which Prize officials have hung a sky-blue banner that reads ">100MPGe." When it's time, the cars proceed through the tunnel into Pit Lane.

If you want to walk from Pit Lane to the track, you first have to climb over two low, white, concrete walls. Now you're standing on a black-and-white checkerboard: the Winner's Circle, where victorious NASCAR drivers get doused with champagne. If you keep walking, across twenty yards of lush grass, you begin to feel yourself climbing a small hill. This is because the track is "banked," meaning its outside edge is higher than its inside edge. The angle of bank at the start/finish line is 12 degrees, and on the turns it's 18 degrees, to help the cars maintain speed while cornering.

The X Prize cars won't be going fast enough to take advantage of the bank the way race cars do, but that doesn't mean they'll have an easy time on the track. Today's highway test, for instance, requires the cars to make 45 laps for 90 miles, maintaining a 45-mile-per-hour average speed and stopping every 10 miles. This is simple enough for a production car, but for an untested prototype driving on a hot day, on a loop of cracked asphalt, 90 miles is a serious threat to the life of the machine.

During the highway test, Team Illuminati gathers in Pit Lane to watch Kevin make left turns. The mood is funereal. At this point, any race could be their last, so they try to observe the run with the requisite dignity. Their car has never been driven this long—or even a tenth this long. Five days ago, their battery pack was dead. Three days ago, they made it through the second Shakedown, but barely, and without successfully completing the 0-to-60 acceleration test. If they can't get the car to accelerate here at Knockout, they'll be sent home.

"It is what it is at this point," says Nate. "The highway test could destroy our battery pack. There's no way to know."

Kevin starts in second position and quickly maneuvers into first, passing the Germans from team TW4XP. The TW4XP, a clever little pod vehicle intended for urban commuters, is one of the smallest cars in the competition. Kevin has made friends with the TW4XP driver, Wolfgang, and now, as Kevin surges past him, he honks his horn twice and blasts "Magic Carpet Ride" through Seven's sound system. A judge cries, "Turn it up!"

A brisk wind blows through Pit Lane. The sky is bright blue and streaked with low-drag wisps of cumulus. Kevin beeps three times as he passes the start line, prompting Jen to mumble, "The horn takes electricity, mister."

A documentary filmmaker standing at the edge of Pit Lane turns to Jen and says, "This must be interesting for you. You're not professional engineers."

"Well, Kevin and George are professional engineers," Jen says. "But not, like, *auto* engineers."

Kevin completes the run without incident. When it's over, he and the Germans drive back to the big white charging tent and hook up their cars to the charging stations. I walk over to the tent to say hello and pass Oliver walking in the other direction, toward the garages. He says he just spoke with the Germans from TW4XP. "Really good guys," he says. "And they talked to me about their figures. They're going to make it, but *barely*. Some of these other teams, they're just dreaming."

MONTHS EARLIER, when Kevin was spraying foam onto the steel frame of Seven, something unexpected happened at one of the joints. A little excrescence formed, a whitecap of excess foam, next to the rear window. The foam hardened into a shape that resembled an alien body, with two arms and an elliptical skull. It was on the inside of the car, behind the backseats. Kevin could see it easily when he looked in the rearview mirror; when he hit the brake lights, the alien form glowed red. Instead of sanding it off, Kevin drew a pair of eyes on the "face" with a black marker, marked an X above the left breast (for X Prize), and gave it a name: Voodoo Steve, in honor of Steve Wesoloski, the head judge. "Voodoo Steve is from the primordial foam," Kevin says. "He watches over us."

Kevin joked with the judges all through Shakedown, and now, at Knockout, they've begun to joke back. You can tell they're glad to have him around. But it's not clear they've ever taken him seriously as an engineer. All along, Kevin and Jen have gotten the sense that they're being humored, valued for the appeal of their story—husband and wife, small-town underdogs—rather than for their minds.

I find Kevin inside the charging tent, slouching in a canvas chair. The four gullwing doors of Seven are flipped up. The Prize is providing electricity via 4-foot-tall charging stations with long black hoses. But it's the responsibility of each team to figure out how to get the electricity into its car. The black hose puts out alternating current at 220 volts, and Illuminati's battery pack requires direct current at a different voltage, so the team has rigged up a charging device—a blue box that's plugged into a green box, which then is plugged into the car—to make the necessary conversions.

Kevin is telling the team the story of his final lap. While running the car in economy mode, which limited power consumption, he decided to try something. Coming around the final corner, he steered the car high up on the banked wall and floored the gas. He got it up to 77 miles per hour. He kept his foot on the gas. He edged up to 83. The Prize speed limit is 70.

"Not many of the cars can do 83 miles per hour," Kevin says with pride.

Nate asks, "If you were not in economy mode, how fast could it go?"

Kevin glances at the car. "I bet it could do what we estimated."

"Two twenty-four?"

"I bet it would hit the high hundreds. 'Cause I was only in third gear."

Jen says, "What if you miss your 67 miles per gallon by just a tiny bit?" In other words, what if that last-lap burst of stupid speed gets them kicked out of the competition?

Kevin just grins.

"Can your grin get any cheesier?" Jen says.

A judge approaches: Jody Nelson, a young engineer experienced in alternative-energy technologies. It was largely Jody who devised the inspection procedures for the electric cars. Having vowed that no electric cars would catch fire on his watch, Jody has set upon them with a torturer's tool kit of voltmeters and probes. He's done his best to break the cars in the garages so they won't break on the track. Here at the Speedway, he's respected and feared in equal measure.

Jody glances at the numbers on the charger hooked up to the Illuminati car. Then the corners of his mouth flicker. It's not subtle at all. A nerd heart is overflowing. It's a look of barely suppressed excitement.

"I think you went over seventy," Jody says.

"No one ever told me the speed limit since we've been here," Kevin says, unconvincingly. Then he shoots Jody a look of fraternal mischief. "You would've, dude." There's no point in denying it. He did the crime and everyone saw; besides, he figures that even the super-strict officials won't begrudge him a little fun.

"No, I wouldn't have," Jody says.

Kevin asks Jody what his MPGe is.

"Start charging really hard after 6 A.M., then you'll find out Thursday," Jody says.

As it turns out, Kevin already knows his number, roughly. He just drove 90 miles and used half of his battery pack. Fully charged, the pack is equal to about 1 gallon of gas. So he got, roughly, 180 miles per gallon. Later in the day, the Prize's Julie Zona, chief liaison to the teams, confirms his guess: 188 MPGe.

"Oh my God, your numbers are unbelievable," Zona tells him.

After news of the 188 MPGe number gets out and spreads

through the garages, Oliver approaches Illuminati in the charging tent and says something. Five hours later, in the lobby bar of their hotel, members of the team are still mulling what he said.

"He's like, 'This makes no sense to me,'" Jen says at the bar, a lock of hair escaping her backwards baseball cap and falling down over her eyebrow.

"'I don't know how you do it,'" Kevin says, quoting Oliver. "'I'm dumbfounded. Your car is heavy. You have all these batteries. I don't mind if I lose, as long as I lose to you.'" Kevin nurses a Parrot Bay mojito. When he drinks, which is almost never, he prefers "girlie drinks."

"Oliver doesn't understand the system," Kevin says.

"We were shooting for the clouds," Nate says, saucer-eyed. "So if we hit the horizon, we're still better than we need to be." Nate flags down a waitress and orders mozzarella sticks. "The car just drove 90 miles straight. She keeps you humble. She makes you do everything two or three times." (For Nate, the car is a "she.")

Illuminati has been poised for so long on the edge of mechanical failure that the idea of winning has rarely come into the picture. It's always been about surviving until the next phase, about not getting left behind. But now the field has narrowed significantly. Only five teams with mainstream cars remain: Illuminati; Edison2; West Philly; BITW Technologies, the Indiana team; and a team from Colorado, American HyPower, competing with a Prius it converted to burn hydrogen fuel in addition to gasoline. (Aptera is still alive, too, in the alternative side-by-side class.)

Five teams, $5 million up for grabs. Tonight, Kevin, Jen, and Nate allow themselves to imagine what it would be like to actually win. And they decide that, for the first time, they're glad to be one of the poorer teams, because it means they have no investors or sponsors to answer to. They're glad they aren't Oliver, for instance. They wouldn't be able to deal with the pressure.

19

Heat

"Fuckin' engineers," says the Aussie mechanic Reg Schmeiss, cracking a smile outside one of the Speedway's garages. "Fuckin' useless. Even worse than before. It's getting hard for all of us. We're all pissed."

It's June 22, the second day of Knockout. Reg says that last night someone realized that the rear wheel pods on one of the Very Light Cars didn't meet the Prize's ground clearance requirement; they were too low to the ground by about an inch and a half. The mechanics had to cut an inch and a half off the bottom of the pods with a power saw.

The mechanics found this damn near intolerable. They hate surprises at all times, but they especially hate them at the track. The track is not the place to be making sudden drastic changes. They blame the engineers. The Very Light Car is too perfect, too "tasty," for its own good.

"The fuckin' $7,000 parking brake," Reg says.

Ron Mathis's parking brake.

"It's beautiful, though," I say.

Bobby says, "It doesn't really matter. It's subjective."

I tell Bobby he's hard on people.

"Everybody starts at one hundred with me," he says, shaking his head. "Then they go down." He gestures toward one of the consultants whom Oliver has hired to help Edison2 with its engines. "That guy started at two hundred, and now he owes me a million. Me and Reg, we don't pretend to be able to do anything we can't do. We're here, here's what we can do, here's what it costs, it's fair, it's forecast, that's it. These guys promise you the moon and give you a moon pie."

ON JUNE 23, the third day of Knockout, a judge waves a green flag. Six cars come off the starting line, all of them virtually silent except for the Very Light Car and the West Philly Focus. The armature of the Focus's electric motor squeaks as it passes. It sounds like an old trolley car.

"That's a defense mechanism," Jacques Wells says.

This is an emotional moment for these teams: their first fuel economy run of the competition. No more major engineering overhauls, no more excuses. Now the cars just have to perform.

West Philly gives a team cheer. Jacques does a backflip across the Winner's Circle.

Meanwhile, Oliver is narrating his emotions. "*Two years,*" he says, tears appearing on his cheeks out of nowhere. "It's so hard. It's so hard. You lose a coil, you lose a plug. You get the wrong flat tire! Two years!"

Jacques does a second backflip.

A group of Aptera employees and company sponsors have segregated themselves in an area off to the side. One guy sits on a golf cart, typing on a laptop. A few of them wear wireless headsets, so that they can speak with their driver, Tom Reichenbach.

Tom, like all the other drivers, has been given a set of instruc-

Two Very Light Cars at Knockout.

tions for this run—an urban drive cycle. He has to make four full
stops per lap. For two laps, he has to accelerate from each stop to
25 miles per hour within a certain amount of time. Then for the
next two laps, he accelerates to 30, and then 35.

It doesn't seem like there's much room for driver creativity.
But Edison2's Ron Mathis has noticed something. There's no rule
against using the natural altitude of the track. Ron has told Brad,
who's driving one of the Very Light Cars in this run, to use the
bank to save energy. He can coast to a stop by steering the car to
the high side of the track. Then, to start again, instead of using
gas, he can just point the car downhill.

Because Edison2 has the most skilled drivers of any team,
they're able to make the best use of little discoveries like this.
Emanuele Pirro has been joking with Oliver and Mathis about
the low speed that he's required to maintain during the economy
runs—"Oh, great," he told them, "a hundred miles behind the pace
car"—but he turns out to have a scientific bent, and he enjoys the

physical and mental challenge of what Oliver calls "driving to the perfect trace." Emanuele keeps a digital stopwatch with him in the car, and he will be able to repeat his lap times almost exactly, to within a fraction of a second. He will figure out the ideal rate of acceleration after each stop, the rate that will use the least amount of energy. He will walk the track before each event, so he can use any imperfections in the track to his advantage. "What the driver does is a very large part of the overall economy picture," Ron tells me. "I think it's a good thing."

Out on the oval, as Brad continues to accumulate urban-cycle laps, he watches the dash readout that tells him the temperature of the water in the cooling system. Ron has decided that any temperature over 135 degrees Celsius (275 degrees Fahrenheit) is worrisome, and exceeding 145 might be fatal to the car. Edison2 doesn't have a headset system to communicate with its drivers, so from inside the car, Brad informs Ron of the water temps by sending text messages. Brad keeps one hand on the wheel and the other on his cell phone. (Brad texts Ron dutifully, because if he doesn't, and something bad happens to the car, Edison2 will blame him. "Basically, it covers my ass if the car blows up," Brad recalls later.) Their conversation looks like this:

Brad: *Water 112*
       114
       127
       What should I do
Ron: *Stop at start finish [the start/finish line] if temp is still rising*

Brad completes a few more laps. Then the sky blackens. Rain suddenly comes down in great sheets.

Eric Cahill, director of the Prize, shouts into his walkie-talkie, "Clear the track, we got hail!"

Everyone runs into the tunnel for shelter as the rain and hail ping loudly against the roof. All at once the cars pit and the drivers pull off their helmets and lope across the grass to the tunnel, rain darkening their clothes. Simon meets his teammates there. Huffing to catch his breath, he says, excitedly, "That's a steep-ass embankment. I caught Aptera several times, and I got to pass them on Pit Road. Which has to be the highlight of my week."

From the mouth of the tunnel, Oliver gazes at the sky. "The cars we have are lightning magnets," he says. "No doubt about it. It's a carbon-fiber body, which is nicely conductive, covered in aluminum foil. Can you think of a better lightning rod?"

AFTER AN hour or so the sun comes out and turns the white concrete walls of the track into solar reflectors. The puddles shrink. The tar cracks and bleeds. It's June in Michigan, and the bronze plaque nailed to the retaining wall of Pit Lane, the one depicting Richard Petty smoking a cigar, grows hot to the touch.

The cars return to the track. One of them is Bobby's car. Oliver has decided to assign a crew chief to each of his four cars. Bobby has drawn the tandem car, the one that got him in the fight with Oliver back in the workshop: one seat in front, one in back. The tandem car has always been kind of an afterthought, the runty sibling of the four. "I have adopted this," Bobby told me earlier, gesturing toward it, "or it's adopted me." The car looks different than it did at Shakedown, mainly because it's now sheathed in a layer of chrome vinyl, in order to reflect the sun and keep the car cooler. It reminds Bobby of a "giant Chipotle burrito." I think it looks more like the skull of a bird dipped in melted metal, or one of the pointy-beaked characters in the *Mad* magazine cartoon *Spy vs. Spy*; when it's whipping around the track, I picture a seam ripper opening the air like a hem.

Bobby's tandem car survives the economy run in good shape.

Then, later in the day, with the sun setting, Brad climbs into the cockpit of the car I've always thought of as the "silver" car: the oldest prototype, the one with the fiberglass body (instead of carbon fiber) and two seats side by side. The green flag goes down. Brad moves slowly off the line. The car makes a noise like a vomiting grizzly bear. Brad immediately pulls off to the side. The engine chugs, and a dozen Edison2 mechanics and engineers run out to the track to see what's wrong. Only Oliver stays behind, chatting with the crowd that has gathered to watch Edison2 struggle to manage heat. "See, none of us have *tested* this shit," he says.

The problem has to do with the placement of the silver car's engine. The body of the Very Light Car is designed for a rear-engine configuration, with the engine placed behind the rear axle. That's where the inlets direct the air: to the rear. But the silver car is different from the other three Very Light Cars. Because of an equipment delay earlier in the year, the mechanics were forced to mount the silver car's engine in *front* of the rear axle, in a "mid-engine" configuration. And now it's too late to move the engine to the rear. Officials, citing the rules, have prevented Edison2 from making the change. What this means is that, out on the track, Brad is driving a misconfigured car—a car with the wrong body for where its engine is.

He manages to get it moving again, but after a few more laps, he checks the numbers and sees that they're edging into the red zone, the place where the car will overheat and shut down completely. The cabin is as warm as a sauna. His clothes are drenched with sweat. He texts Ron the water temperature: *145.*

Ron knows the engine is about to destroy itself. But there's nothing he can do. There are no do-overs at this point. If a car doesn't complete an economy run, it's eliminated. He texts back: *Drive it until it won't move anymore. Ignore all dash warnings. Only stop if you smell/see coolant.*

Brad steers up to the top of the wall. The engine cuts out. He goes forward 40 yards and then stops. Bobby and the rest of Edison2 run across the grass to the car.

There's a firewall that separates the passenger compartment from the engine. Edison2 needs to get air to the engine so it won't blow up, and to do that they have to rip off the access cover to the firewall—a piece of metal. Bobby reaches for the metal. It burns his hand. He tries again, lifts it off.

Now the heat from the engine bay rushes into the passenger compartment, bringing the temperature close to 140 degrees Fahrenheit. (The rules don't require traditional air conditioners, so the Very Light Car doesn't have one.) Brad's next few laps are the hottest he's ever experienced in a car. It isn't long before the extreme heat overwhelms the engine and the car comes to a stop. Brad gets out and tries to take his helmet off, but he can't, because the plastic is so hot. The X Prize decals on the side of the car are shriveling up, falling off. Brad tosses the key to a mechanic, and the mechanic quickly drops it, to keep it from burning his hand.

It turns out that the engine's throttle body, a part that controls air intake, has failed. The throttle body is one of the few parts that Edison2 didn't design from scratch. The team bought it off the shelf. "You're stretching the systems to the end," Brad tells me later. "When you do that you're setting yourself up for different failures. I think maybe we're spreading ourselves too thin."

Oliver explains the failure of the throttle body to me by asking if I know the old joke about the farmer who's upset that his horse died. "He was just about to pay off," the farmer says. "I was feeding him less and less each day, and I'd only just gotten him down to eating nothing." By design, the Very Light Cars are starved of fuel, and now they appear to be starved of air as well. Oliver says, "We're going to tame the racehorses and turn them into docile workhorses. It just takes time."

LATER THAT night, working in the garage, West Philly's Simon Hauger starts to feel anxious.

The Harley, he thinks. *We're out of time.*

Tomorrow, Simon has to drive the Ford Focus hybrid in the highway test—45 laps, 90 miles. The car's electric battery pack can't generate enough power to last all 90 miles. Simon will have to use the gasoline side of the car—the Harley. There are two problems with that. One is that the Harley is dramatically less efficient than the electric motor, which means that the more he has to use the Harley, the lower the car's MPGe will be, and Simon is already worried about efficiency. In other competitions, such as the Tour de Sol, West Philly's hybrids have achieved upward of 180 MPG, but that was with a lot of cruising at highway speeds. The drive cycles in the X Prize are a lot tougher, because they replicate the way normal people actually drive. Simon thinks the Focus will get between 100 and 120 MPGe and hopes he's not off by a factor of 2.

The other problem with needing to rely on the Harley is that it simply might not rev up. It's always been the least reliable component of the car's design.

Simon shares his fear with Keith Sevcik, who's up late, pouring new, untested code into the Focus's electronic brain—code that will allow Simon to switch between the car's electric motor and the Harley at the push of a button. And something suddenly dawns on Simon: *It's either going to work or it's not.* There's nothing else they can do at this point, so why not go out and have fun and see what happens.

The next day, the fourth day of Knockout, Simon settles into the driver's seat of the Focus wearing jeans, a blue pullover, and the crash helmet that, according to his wife, Cindy, who is here today with their two young boys, makes him look like Speed Racer's geekier older brother. The West Philly students push him onto the oval. (The teams push their cars instead of driving them because they don't want to waste fuel.) It takes six kids and adults to push the Focus. It takes just one man to push one of the Edison2 Very Light Cars. Simon lines up with several other cars, including the Very Light Car, the Aptera 2e, the E-Tracer covered motorcycle, and the Zap Alias. The driver of the Alias is IndyCar champion Al Unser, Jr., also serving as a celebrity spokesperson.

Simon starts the test in electric mode and passes the Very Light Car, which pits right away. There's a tumble of gray as nine men in Edison2 uniforms sprint from Pit Lane to the apron of the track. Simon watches the spectacle out of the driver's-side window. It looks like another overheat.

After the fourth or fifth lap, Simon decides he has no choice. In third gear, going about 45, he presses the button to start the Harley.

Meanwhile, over in Pit Lane, Keith cranes his neck to make out a whitish dot coming around Turn 3. He thinks he hears a familiar noise.

"Is that the Harley?"

"Wow, that is strange," says a journalist from *Automobile* magazine. "I've never heard a sound like that coming from a car before."

"It hasn't been road-tested with the new code," Keith says. "Here it comes." Keith grins and slaps the wall with his hand. "It's working." He darts twenty yards down the track, screaming, "It's working! It's working!" When he returns to his previous viewing position, Oliver is standing there.

"I have to take my hat off to you for building two completely different cars," Oliver tells Keith. "If you want to be technical, at this moment you are ahead of us. You have two for two, we have three of four. But enough. Hat is off. And you beat Cornell!"

INSIDE THE Focus, Simon notices two things at about the same time.

One: *The Harley runs better when I go faster.*

Two: *That is Al Unser, Jr., up ahead of me.*

In fourth gear, he lets out the throttle and maneuvers to the high side of the oval, creeping up on Al Jr.'s Zap until he's pretty sure Al Jr. can see him in his rearview mirror. Then, at 58 mph, the former high-school math teacher passes the two-time winner of the Indianapolis 500.

As Simon accumulates laps, the Harley gives him no problems, except that its vibration heightens his need to go to the bathroom. After the 45th lap, Simon pulls onto the grass on the far side of the finish line. The only car to finish before him is the E-Tracer. The kids rush across the grass to greet Simon as he emerges from the car, his cheeks shiny with sweat.

Keith jumps up to hug him.

"Ha ha ha!" Simon says.

"So, Simon," asks Ann Cohen, "did you see Edison2 at the beginning?"

"It was like déjà vu all over again," Simon says.

There's a commotion in a patch of grass next to the West Philly team. Aptera Motors has come in right behind Simon. Before the test, the Aptera engineers weren't sure if their battery pack would last for all 90 miles, but they made it with juice to spare. They celebrate by popping a bottle of champagne.

"We did it!" says Aptera engineer Brian Gallagher, looking more relieved than excited.

Tom Reichenbach, dressed in a white jumpsuit, smiles and says, "High highs and low lows."

"This makes up for all of it," Brian says. "We were tellin' Tom, 'You can pass a guy if you want, you have the energy,' but he played it conservative. Classy, I guess."

Cindy Hauger says to Ann, "Did you see what the driver of Aptera wears versus what Simon wears?"

Someone on the West Philly team asks Simon how many days there are from the end of Knockout to the start of Finals.

"Ten?" he says.

More like three weeks. Still, not a lot of time.

LATER THAT evening, Simon, Ann, and the other adults throw a pizza party for the kids by the pool of the hotel where they're all staying. I eat a slice and chat with Ann. She mentions that, earlier in the day, she had a brief conversation with Edison2 aerodynamicist Barnaby Wainfan. "Barnaby didn't seem to understand what I was telling him," Ann says. "I was telling him that there's just no way my ass fits in the Very Light Car."

After the party, she goes back to her hotel room. She's not sure what Simon is up to. Someone says he's with Keith in the hotel, crunching numbers. Ann can't get to sleep.

At 11:18 P.M., Simon sends her an e-mail:

> Ann,
>
> So it looks like our fuel efficiency is a little low on both vehicles. We'll have to wait and see what the official numbers are tomorrow.
>
> Simon

The message ends, as do all of Simon's e-mails, with a quote from Aristotle: "All who have meditated on the art of governing mankind have been convinced that the fate of empires depends on the education of youth."

Ann reads the e-mail again.

> Our fuel efficiency is a little low.

What does that mean, "a little low"? Low as in 80 or 90 MPGe—below 100 but more than enough to advance to the Finals? Or low as in below the 67 MPGe cutoff?

She shoots back three e-mails.

One: Are they calculating fuel efficiency tomorrow?

Two: This is one of those cases where we could

         get it just right ... next month ... August?

Three: How low?

Ann waits.

At 11:43, Simon finally replies:

gt—59

focus—64

We're close and there are LOTS of variables but if
you've been waiting to pray, now is the time.

20

Goonies Never Say Die

n June 25, the fifth day of Knockout, Simon arrives at the track before 8 A.M. and goes straight to the charging tent to check on the cars. He sees a judge approaching. The judge is carrying a blue binder. The binder contains the official economy numbers of both West Philly cars.

Simon has a sense that the numbers aren't going to be good, and not just because he and Keith did the math. Last night at the track, Simon tried to ply one of the friendlier officials, Julie Zona, for information.

"Do you have any official numbers?" he asked, trying to sound nonchalant. "We did our math and we were kind of short, and we're anxious to get the numbers."

"Well," Zona said, "math doesn't lie."

Shit.

Simon now opens the binder.

GT: 53.7 MPGe

Focus: 63.5 MPGe

So Simon's math was roughly correct. The GT isn't close to the 67 MPGe cutoff. The Focus is achingly close—about 5 percent away.

Something doesn't seem right. Sure, maybe the cars are too ordinary to ever reach 100 MPGe—Simon can accept that—but fifty-four? sixty-four? How could his estimates be off by such a large amount?

He communes with Keith, who has the idea to plug his laptop into the computer system that controls the Focus and download a file in which voltages, temperatures, and currents have been logged—the complete electrical history of the car. Crunching the numbers, Keith discovers something troubling. The officials made their calculations based on the total amount of energy that the cars drew from the charging stations. But only a portion of that electricity made it into the Focus's batteries. Keith now realizes that West Philly's battery charger—the contraption that feeds current from the charging station into the car's batteries—is faulty. No one on the team ever noticed. They built it at the last minute because it's such a simple, minor piece of equipment, and they were concentrating on solving harder problems. Yet the charger has probably cost West Philly a chance to get to Finals. If it had been working correctly, they could have gotten the Focus to about 80 MPGe, well past the cutoff.

Simon breaks the news to the adults and the few kids who arrived early. "We're proud of you," he says. "Something was wrong with our charger, but our work was outstanding."

At least two of the adults start to cry. Jacques Wells takes a long walk along Pit Lane and comes back to the garage energized. "We're still one of the greatest competition teams in the industry," he says. "Listen, if we had half of Aptera's budget. *Half.* Come on now."

Keith cracks open a bag of chips and eats them with a dazed look. "Now I have to find a job," he says.

More kids arrive at the Speedway. They haven't heard yet, so Simon gathers them in a circle for a speech. "The reality is this," he says. "We've built cars that have gotten over 100 miles per gallon. We've never driven on a cycle this aggressive. All the cars are having a hard time." The issue, he says, wasn't the quality of their work. It was "the *design* of the car." They'd chosen to adapt an off-the-shelf vehicle, which limited their ultimate fuel efficiency.

Ann jumps in. "We went out on this track and got 65 miles to the gallon in a frickin' 3,000-pound Ford Focus," she says. "Come on. That's unbelievable."

WEST PHILLY isn't the only team processing bad news. George Voll of BITW Technologies has also been knocked out. George runs a hardware store. His car, a biodiesel-converted Chevy Metro, clocked in at 51.1 MPGe. "The city cycle is probably what killed us," George says glumly. "I'm still glad we made it so far."

Nearby, the American HyPower team, from Colorado, is packing equipment into a van. Their Prius conversion got 54.5 MPGe, about the same as a stock Prius. Team leader Mills Snowden says he thinks no one will win the mainstream class. "There are only two teams left," he says. "Of those, Edison2 can't make the emissions, and Illuminati can't go the distance" in the range test. Disgusted, he drives away.

Three years, eighty cars, sixty-seven teams, millions of dollars spent, hundreds of credit cards maxed out, and the mainstream class has come down to just three cars and two teams: Edison2 and Illuminati. The racers in the jeans factory, the hackers in the cornfield. The South and the Midwest. Internal combustion and electric. No one ever picked either team to win.

Late that morning, under a blinding sun, one of Edison2's mainstream cars and Illuminati share the track for the range test. To advance to the Finals, each car has to drive for at least 134 miles, at an average speed of 55 miles per hour—the highest speed of any event in the competition. Oliver leans against the concrete retaining wall and watches his last remaining competitor in the mainstream class zip past. "That thing is surprisingly light for how heavy it is," he says of the Illuminati car. "Illuminati has a few more mountains to climb. I don't think they've done the acceleration yet. But I tell you what. I respect them.

"I don't even want to stand here," Oliver says. "It's too nerve-wracking. I'll pop."

His own car is pulling over and stopping every five laps. There isn't enough air getting to the radiator—something might be wrong with the inlets, something that's only showing up now due to the car's relatively high speed—and the engine's overheating. Each time the car pits, Ron throws a bucket of water into the back of the car to cool down the engine. Then Ron pushes it for a few yards until the engine kicks back in. Racers often talk about pushing cars to the "ragged edge," to their absolute physical limits, and now Oliver says, "This car is so on the edge."

He gets on the phone to David Brown back in the garage. "We need antifreeze," he says. "We have no antifreeze. Find it from the pits, find it from other teams. *Find it.*"

Given the required pace, Edison2's car can only afford about fifteen minutes of total pit time. It has already pitted several times, eating up precious minutes. Ron isn't sure if they'll make it in time.

The car comes back around to the starting line and pulls off to the side. Ten Edison2 members are waiting for it on the edge of the grass. They empty water jugs on the back of the car; steam billows off the scalding metal. P.K., Edison2's crew chief and senior mechanic, counts down: "Five seconds. Four. Three. Two. One. Go."

Barnaby Wainfan says, "Until you finish, it ain't real. It's a hope backed up by analysis."

P.K. takes a drag on his cigarette and walks over to talk to Barnaby.

"Just enjoying the excitement," P.K. says. "The *drama*." He smirks. "Oh, I've been well entertained. Take that to mean what you will."

"Oh, we've had some moments," Barnaby says.

P.K. shoots Barnaby a look of mock horror. "Not you and me? Well, maybe. Maybe a couple."

The aerodynamicist looks nervously at the track.

"I hope none of the changes have made it worse," he says. Barnaby means changes like the one that raised the car's ground clearance by an inch and a half—quick-and-dirty fixes made by the mechanics here at the track. Barnaby is worried the car's aerodynamic sleekness has been compromised.

"Well, you have to try *something*," P.K. says calmly. "To make it right. Under duress."

"Oh, I understand that," Barnaby says. "Everybody's doing their level best. But it's hard not to chew on."

OLIVER CAN'T bear it any longer. He needs a distraction. He walks over to the West Philly garage bay. He finds Simon there and shakes his hand. "You did something special," Oliver says, and then launches into a mini-autobiography. When he was fourteen, he says, he didn't have good grades. His first SAT score was 270. But he had a shop teacher, a man who believed in him, who taught him to appreciate automotive beauty, and by the time he graduated from college, he owned a Ferrari. "Thank you," Oliver says to Simon, "for being one of the shop teachers who does the best job in the world."

There's something classically Oliver about this conversation: He presents himself as the mayor of the competition—the sweet, avuncular, encouraging father figure—at the same time that he glorifies himself. His face reddens. A tear streams down his right cheek.

"I don't know why I'm crying!" he says. "You did a shitload

of work, and now you're here. It's an achievement. It really is. Ann might have a similar story. She is crying too."

"I am," says Ann, wiping away a tear.

Jacques Wells stands there, arms crossed, not sure what to make of this large man in distress. Oliver's eyes meet his. Oliver wobbles over and bear-hugs him. Jacques pats him on the back like he would a baby. (Jacques will later say of Oliver, "I love that guy. He's hilarious.")

Oliver detaches himself. "I've got three hundred years of stuff to get done and I've only got twenty-five years," he says. "My knees hurt. My legs hurt. Like the time clock's *on*. . . . We gotta make something. We are fixing stuff our parents made. . . . I don't want to sound like a social radical . . ."

Simon responds, "I've been through four different reforms in the school system and they're all the same. Sit the kids in desks and beat them to get results. Any business that ran that way, you'd be bankrupt—50 percent of your product is a failure."

It seems like they're talking past each other, Simon and Oliver, trying to share their most cherished beliefs all at once in sound-bite form. Oliver abruptly shifts topics. "I will be honest," he says. "I am still not a huge electric car booster. I am quite shocked at how well a lot of them are performing. I figured that a lot of them would fall off a cliff. But that proves Mr. Diamandis right, which is that if you get a lot of guys banging their head against the wall, you get somewhere." He adds, "Between us and Illuminati, it *could* be a race. Because they have demonstrated that their car can handle. We have demonstrated that our cars stall. It's really crazy! I would have never thought it. But it could be a race!" He twists his face into a knot. "Those guys, they may be from the country, but they are not . . ."

He stops himself. "Visit me if you ever want to. I gotta go." Then he walks out of the garage and back to Pit Lane, where, much to his surprise, the Aptera 2e sits motionless by the side of the track.

Barnaby says the car overheated. "Looks marginal," he says of the 2e. "Very slow. They melted down doing 67 miles on a smooth day on a good track with no hills." Remembering that the Very Light Car is also overheating, he adds, "Part of cooling is aerodynamics. There's a part of me saying, what did I miss? I don't want to be the guy who costs Edison2 their Prize."

The Aptera 2e fails to finish its final lap. The car has to be pushed back to the charging tent.

THAT DAY, Friday, while Kevin Smith of Illuminati Motor Works is doing laps around the oval on a long economy run, he grabs his phone, dials his wife, Jen, and jams the phone into his crash helmet so he can keep his hands on the steering wheel. "Hi, hon," he says, "it's me. I just wanted to tell you I love you. I can't hear you right now. I've got the phone shoved up inside my helmet. I'm on the track. I'm going to repeat it right now because I can't hear you. Hi, hon, it's me, I love you."

Jen is back in Springfield, Illinois. She went home early, to save herself the stress, and because she wasn't sure how she could be helpful. All day she's been following the Prize on her computer. In an attempt to engage the public, Prize officials have set up a live video feed from the track, along with a chat room moderated by an employee who has been typing periodic updates on how the cars are doing. Jen stays riveted to the video feed, but she can't figure out what's going on with Aptera. She saw the 2e break down and fail to complete the final lap of its run. Why wasn't there an announcement that Aptera had been eliminated? Was Aptera Too Big to Fail?

The following Monday, in Michigan, Kevin's mood verges on euphoria. The car is working great, doing everything it's supposed to. By now he has completed the braking test, the Moose Test, and all three drive cycles, with only the acceleration and range tests yet to come. He's so close to the end. And everyone else is dropping away. For the first time in the competition, events seem to be lining up to help him instead of beat him down. Even Oliver is starting to talk to him, to say nice things—buttering him up, almost. *He sees us as a threat now*, Kevin thinks. The ultimate compliment.

Hell, we could win this thing.

In the early afternoon, Kevin drives to one end of Pit Lane and comes to a stop at a cone. It's time for his acceleration test. He needs to reach 60 miles per hour within 15 seconds. The car has never been able to do this, thanks to its temperamental, perpetually slipping clutch—the car's weak link. Kevin has white-knuckled it through the other tests in a single gear, but to hit the acceleration target, he needs to shift from first gear to third rapidly, taxing the clutch.

A few seconds into his first attempt, he tries to switch from first gear to second. The clutch slips. It feels like he's trying to run up a hill coated with wet leaves. More than 20 seconds tick by before the car breaks 60.

Kevin circles back to the starting line and tries again, then again. On the next few attempts, he uses different combinations of gears and feathering techniques, letting his foot on and off the clutch, on and off, on and off—double-clutching, triple-clutching—but each time the clutch refuses to grab.

He stops the car. Nate is there.

"Nate, this is all it's got, man," Kevin says. "It's not getting any faster. Our clutch just went poop."

"We're fucked," Nate says.

Kevin still has several more attempts until he reaches his limit of 10, but he knows it doesn't matter. He suddenly feels violently ill. Every muscle aches. He wants to lie down on the asphalt and sleep for a year. But that isn't the worst part.

Oh, shit, now I have to tell the other guys.

They're there on the sideline, on the edge of Pit Lane, waiting and hoping: Nate, George, Kevin's father.

Well, maybe I can try smoking the clutch. Heat it up a little. Maybe it'll catch. Or if that doesn't work, maybe I'll let the batteries rest, let it run cooler. Fuck it. If we're going home, we're going home broken. . . .

The next couple of runs, Kevin smokes the hell out of the clutch. Even Kevin's nearly deaf father can hear it screaming in agony.

Kevin asks the judges if he can go off and just practice shifting for a couple of minutes. They say okay. And that's when it happens. He pulls hard on the shifter knob and something snaps. The knob bobs around sickly, like a straw in a lidless drink. Kevin coasts to a stop. *Aw, crap*, he thinks, *I broke the shifter linkage*, the cable that goes from the shifter to the transmission. He's out of gear and stuck that way.

Kevin calls out to Nate, who brings over a pair of vise grips and tries to muscle the car into gear. It works—except now the car is stuck in fourth gear.

Kevin takes the car back to the track. He makes two more attempts to hit the acceleration time. But the car is too slow in fourth gear.

And that's it.

Illuminati has one car. It's broken.

Kevin thinks briefly about filing an appeal. Throughout the day he's been watching other teams go through the appeals process, and none have been successful. Besides, how can he appeal a broken shifter linkage and a junked clutch?

Kevin packs up his things. The car is so hobbled, it won't even pull itself up the ramp of his trailer, so, with Nate's help, Kevin pushes it up the ramp. He says goodbye to the judges and the other teams. Julie Zona hugs him. Oliver says, "I wish you hadn't gone out this way. I wanted to race you."

Jen calls, but she doesn't reach him. Kevin's phone is off.

———

NO ONE is sure what's going on with Aptera. Are they in? Are they out? The Prize makes no announcement.

The longer the silence lasts, the angrier the other competitors get. It's not like the officials have been shy to pull the trigger on teams that suffered mechanical failures. An Italian team blew a piston in its gasoline-powered prototype; a Canadian team competing with a three-wheeled hybrid couldn't complete the emissions test. Both teams were sent home.

In a discussion thread on the Prize website, anonymous commenters start asking pointed questions:

> charly wrote: IS THIS A JOKE? So who's righting the
> rules at X Prize? Why can't they enforce the very
> rules they write? . . . is this contest rigged what's
> the point of this hole thing
>
> ZippyZipZap wrote: charly, that is right! . . . If a team's
> vehicle is still having problems or is not ready
> for the competition, they should be eliminated,
> regardless of its public/political standing. Period!
> What's next? Are we now going to have separate
> "make-up" Indy-500's for all the cars in that
> competition that were late or could not complete
> the original race? Boo-hoo-hoo.

It isn't until a few days later that the Prize announces its decision: Aptera can still compete.

Prize director Eric Cahill explains the reasoning to me later. It's true that Aptera broke down before completing its run. But it really broke down on only the final lap. And the final lap was one of two extra penalty laps that Aptera had incurred for going too slow. It went too slow because the car overheated and Aptera couldn't fix it fast enough to come in under the maximum lap time set by the Prize: 3 minutes and 10 seconds. But Aptera could have fixed the car more quickly if it only had imitated something Edison2 was doing in that same run. Edison2 was also having mechanical trouble with its car, but instead of waiting behind the concrete wall on Pit Lane and then sprinting across the grass to the track when its car broke down, Oliver was pre-positioning his pit crew on the grass, saving the team precious time (and penalties). Aptera didn't catch on to this trick of pre-positioning its pit crew until later in the race.

Cahill's argument is complicated and technical, but it boils down to fairness. The Prize didn't think it was fair to eliminate Aptera simply because the team wasn't thinking quickly enough to figure out that one of its competitors had taken an advantage. In other words, Oliver's aggressiveness saved his competitor from certain doom. "Edison2 kind of set the precedent, if you will," Cahill tells me.

WHEN KEVIN and his teammates get back to Divernon after Knockout, Jen and Josh are waiting. Kevin is wearing a T-shirt that says GOONIES NEVER SAY DIE, a reference to one of his and Jen's favorite movies: A group of rejects from the poor side of town set off in search of pirate treasure in order to save their families' homes from foreclosure.

"Even before our car was off the trailer," Josh recalls later, "we were already talking about what we were going to do. Kevin puts on a strong face, because that's who he is. There's always tomorrow with Kevin. And that's really rubbed off on the rest of us, too."

Kevin, see, has been thinking.

There's only one team left in the mainstream class. The team's cars are overheating at every turn. Oliver will still have to make it through Finals to collect the $5 million. What if his cars self-destruct? What if he falls short of 100 MPGe? What if Oliver fails?

What will happen then?

What will the X Prize Foundation do? *Not* award a prize? Would that really be in keeping with the spirit of the competition? Isn't there a pretty strong possibility the Foundation will reopen the competition?

"This is all part of the plan, guys," Kevin tells the rest of the team. "We always play it tight. We always play it down to the minute." Maybe breaking their car was the best thing that could have happened, he says. Because now they can make any changes they want. They can replace the clutch. Hell, they can replace the whole damn transmission. They roll up their sleeves, they erase the whiteboard, and they come up with new stuff. Goonies never say die.

And in August, if the Foundation chooses to reopen the Prize as a mainstream-class-only event, they just drop off their paperwork and say, "Hi, guys, remember us?"

part three

the ragged edge

The Edge . . . there is no honest way to explain it because the only people who really know where it is are the ones who have gone over.

—Hunter S. Thompson

21

The Greatest Place on Earth

liver makes it through Knockout as the last man standing. He's now the only competitor left in the $5 million mainstream class and the only one in the entire field with an internal-combustion-engine car. "This event is very much pushing for electric cars, and we are the Antichrist," he tells me after the final day of Knockout. "But we are also a reminder of the truth." He says he feels bad for Kevin of Illuminati—"a really good guy"—but that the X Prize has proven that even talented builders of electric cars have a tough time dealing with the limitations of the technology. Oliver can read it on people's faces and in their body language: They're starting to believe in what he's doing. One day during the event, a team from Magna International, a billion-dollar company that makes parts for the Big Three automakers as well as several European automakers, came to the track to talk with him. "They see us, they spend three hours, they get back in their cars and they leave," Oliver says. "They don't say hi to anybody else."

In terms of performance, Oliver's cars still aren't where they need to be. He survived Knockout with three cars out of the original four: two mainstream cars and Bobby's tandem car. The MPGe of one of the two mainstream cars was measured to be 101.4, but penalties related to the car's cooling problems brought its official number down to 67.3 MPGe. The other mainstream car attained only 80.3 MPGe, while Bobby's car performed slightly better at 97 MPGe. Oliver's still not certain he can crack 100.

"It's just not a sure thing," Oliver says. "We can still break a car. If I had two more months to work on 'em, they'd be bulletproof, you can drive them to Los Angeles." But he has only three weeks until the start of Finals. He could use some insurance, a little bit of margin. Which is why, in early July, one week after the end of Knockout, he brings his team to the GM Technical Center in Warren, Michigan.

Twenty thousand people work at this place, but today, a Thursday, there are few signs of life. It feels eerily deserted. The occasional late-model Chevy threads through the streets like an apocalypse survivor.

It's been a little more than a year since GM declared bankruptcy and began to receive government bailout funds. Oliver, Ron, Brad, Barnaby, and James Nero, an engineer from the company that provided Edison2 with the computer that controls the Very Light Car's engine, drive through the three-hundred-acre campus, gawking. They pass mid-century modern buildings made of colorful glazed brick and a perfectly rectangular artificial pond featuring a gleaming modernist water tower, its bulb reminiscent of a UFO. When construction began on this place in 1949, newspapers called it the Versailles of Industry, a monument to a particularly American vision of strength and progress. "In these laboratories, as in others," GM's director of research said at the time, "scientists are at work extending the frontiers of our knowledge of the physical universe." He argued that major investments in science and engineering were crucial if we were to defeat Communism: "We are in a merciless struggle for survival, freedom against regimentation, self-discipline against coercion." The Center still feels haunted by that ambition and idealism; it's like entering an old *Popular Science* spread about the future, perfectly preserved.

Oliver and crew eventually reach their destination, a sprawling rectangular building on the far end of the campus—GM's main automotive wind tunnel. I arrive at the tunnel not long afterward. A young woman comes out to meet me: Suzy Cody, a GM aerodynamics engineer. Her black hair, tied in a ponytail, is streaked with blue. Cody leads me to a garage beneath the main part of the tunnel, where I see the men of Edison2 gathered around one of the Very Light Cars. Brad and James brought the car here yesterday. Next to the Very Light Car is a Buick Lacrosse.

Cody suggests that, while we're down here, preparing for today's wind-tunnel tests, we might want to check out the tunnel's most important instrument—its giant, ultrasensitive balance.

A GM technician with a white mustache and a weary look offers to show us. He leads Oliver, Ron, Barnaby, and me into what feels, at first, like a boiler room.

"Well, here she is," the technician says, with the slightly bored, resentful air of a man who lives next door to the Pyramids and doesn't get what all the fuss is about.

"She?" I ask.

The man lowers his head. "Well, she's colorful."

The balance is basically a big pile of metal that stretches up to the ceiling, twenty feet high—a massive and exquisitely sensitive version of the scale in your bathroom. It measures with great accuracy the forces on the test vehicle, allowing engineers to determine the vehicle's coefficient of drag as well as other important quantities. The balance has its own concrete foundation, separate from the building's, to ensure that it won't vibrate. The GM man becomes momentarily animated. "The car goes up there, on the red thing," he says, pointing to the uppermost layer of metal, which is painted red. "Then the red thing pushes down on the yellow thing, which pushes down on the green thing. You can drop a quarter on the bumper, and this machine will know it. Yeah, this thing is bad to the bone, bad to the bone."

The machine dates back to 1978, he says, when the tunnel was built in response to Congress's passage of national fuel-economy standards. GM knew it had to make more efficient cars, and improving the aerodynamics of its fleet was one way to do it. Something in the technician's voice changes when he says 1978. He pronounces it with solemnity, as if he were speaking of a time far in the past, perhaps hundreds of years ago. "Back when they actually *built* things," he says.

"I *love* seeing stuff like this," Barnaby says.

Ron takes out his camera and snaps several pictures.

Oliver circles the balance, murmuring softly.

He's been looking forward to this day for weeks. Back in May, when he brought the Very Light Car to the Technical Center to show GM's engineers, Oliver asked if the company would make its automotive wind tunnel available to him. When GM said yes, he was thrilled, because he knew how valuable a day in a wind tunnel would be to Edison2. For one thing, the data generated by the tunnel would show them how to improve the car. But perhaps more important, the data would give Oliver ammunition to use with a potential investor or skeptical journalist who saw his salesman's smile and assumed he was just making stuff up. If General Motors, the largest car company in the world, vouched that the Very Light Car possessed certain physical characteristics, who would be able to argue?

GM is getting something out of the deal, too, which is why they're waiving the usual $2,150-an-hour fee to rent the wind tunnel. One of the main functions of the tunnel is to keep tabs on competitors; GM regularly tests autos from other car companies to see if they're really as aerodynamic as claimed. The Very Light Car represents a whole new category of potential competition. It's a design GM has never seen before, and today the company can get an early peek.

In the tunnel's control room, one floor above the balance, Cody suggests that it might be fun if everyone tries to guess the car's coefficient of drag. The closest guesser will get a case of beer.

Oliver goes first, guessing .148. This is optimistic. The lowest Cd on any production car in history is .195, attained by the GM EV-1, tested in this very tunnel. Cody writes .148 on a large whiteboard in the control room.

A senior GM engineer, Frank Meinert, shifts his weight from foot to foot. He's the one who has been researching the history of the balance; unlike the lower-level GM tech who spoke of "the red thing" and "the yellow thing," Meinert knows all about the wind tunnel's engineering marvels. He has a dry, precise manner, and the other engineers call him Spock. Spock guesses .166. "I don't see things like this very often," he says.

Oliver grins. "You gotta understand, I am the salesman," he says. "My number *has* to be low."

Cody writes down several more guesses from Edison2 team members, which range from .140 (Ron) to .170 (James). She guesses .187, the highest so far. Brad says, in mock horror, "Suzy, what are you *doing*?"

"Hmmmm," says Barnaby, stroking his beard.

To some extent, the poll reflects how much faith people have in Barnaby's skill. He designed the aero of the Very Light Car by eye. Has he gotten it right? It certainly looks sleek, but cars that look sleek are often secretly blocky and are revealed as such by wind tunnels. Yes, the car achieved 100 MPGe before penalties at the Knockout stage, so, as Barnaby now insists, "It can't be *that* bad." Then again, during Knockout, some of the cars overheated, a potential indicator of screwy aero.

"I'll be an optimist," says Greg Fadler, a rosy-cheeked man built like a high-school wrestling coach. Fadler is the executive in charge of the wind tunnel. Not only that, he's the head of all aerodynamics at GM. Fadler doesn't really need to be here. Today happens to be a union holiday, which is why the campus appears so empty. Most GM employees have the day off. But the wind tunnel operates five days a week, three shifts a day; when other parts of the company shut down, the wind stays on. Anyway, Fadler is curious about the Very Light Car. He wants to see it in action.

"Put me down for .190," he says. The highest of all the guesses.

The only guess left is Barnaby's.

"And in the great tradition, the aerodynamics guy has to be the most optimistic," Barnaby says with a flourish. "Put me down for .125."

THE TUNNEL, the largest of its kind in the world, is a strange, epic piece of architecture. It's a big, rectangular tube, with a perimeter as long as two and a half football fields. The air rushes around and around the rectangle in a clockwise direction. Lodged in one of the long sides of the rectangle, there's a forty-three-foot-tall fan made of Sitka spruce. The fan blows the air around and around the perimeter at speeds up to 138 miles per hour. Each time the air reaches a corner, it hits a set of diagonal "turning vanes," steel blades that force the air to make a 90-degree turn.

The car sits in the long straightaway opposite the fan, in a slice of the tunnel known as the "test section." Made of blue-painted brick, and accessible through a door in the control room, the test section feels otherworldly, like an ambush spot in a video game: a cavern lined with gray sound-dampening fabric. (The fabric is necessary because engineers use the tunnel to measure cabin noise.) When you walk into the test section, you're at the center of a long corridor that stretches away from you and slopes downward in both directions. Each end is as black as outer space.

The car rests on a large, silver disc. Beneath the disc, hidden from view, is the balance we just saw, which measures all the nuances of how the car reacts to the air. More than three decades' worth of GM models have sat on this disc—legions of Corvettes and Camaros and Fieros and Bonnevilles and Escalades. And now the Very Light Car is anchored there as well.

As the people of GM and Edison2 wait for the fan to start blowing, Oliver chats up Greg Fadler. "One of the only people we actually want to impress is people like you," Oliver says. "We don't really care about the rest." Oliver's big hope is that Fadler will be so amazed by the Very Light Car's performance today that he'll show the data to his bosses, and maybe one of those bosses will give Oliver a call and begin a negotiation to buy his technology.

"We were enthusiastic from the beginning," Fadler tells Oliver. "For us, it's a unique shape. We can't seem to get traction inside the company to do things like this. You might just be the motivation we need."

A noise rushes up from the guts of the building, a massive sucked-in breath, and then the wind comes: 75 tons of air—as much weight as ten African elephants—speeding toward the Very Light Car. If all goes well, the car will split the air like a shark fin splitting water.

The Edison2 team watches from the tunnel's control room, where two technicians monitor a large black panel beneath a bank of screens that show various parts of the tunnel. The car appears on one of the monitors. A placard next to it says:

GM 19829
RUN NO: 01
VEHICLE: EDISON2 VLC
08JUL10

The only visible sign of the wind on the monitor is a slight flutter, a ripple where the air slides off the back of the car. There's no undulating ribbon of smoke like you see in car ads. Although smoke trails can be useful in identifying areas where the air separates from the body, GM largely uses smoke for publicity. The point of the wind tunnel is to generate data, and the engineers strain to read the small type streaming onto a nearby console.

There are rows of digits for forces, airflow velocities, pressures, temperatures, wind noise, and humidity. In one of the rows, it says: Cd. 171.

Brad, closest to the console, whips his head around. "One-seven-one," he says. James Nero wins the case of beer.

Barnaby, for once, is quiet. He tells me later he's disappointed with this number. He expected better. He watches Oliver's face. Oliver boosts himself into the air by pushing off Ron's shoulder.

"You cannot *imagine* how happy," Oliver says. One-seven-one. Not as spectacular as he hoped, but still a record. "This is one of the greatest tools you can use," he blurts. "It's probably a $50 million machine! I don't know. It's so expensive I can't even imagine." He turns to Fadler. "Point-one-seven-one? Have you ever had anything in that territory?"

"No," says Fadler, arms crossed.

"So if we massage this and get it down to a .140, it's interesting to you, then?"

"Oh yeah." Fadler pauses. "But .140 will be a stretch."

"It's a platform from which to design," Oliver says.

"Sure sure sure. Sure."

It's hard for Oliver to get a read on Fadler. Is he impressed? Jealous? Irritated? Is he just humoring Oliver?

The wind is shut down, and everyone moves from the control room into the test section, which seems to vibrate like a tuning fork. It's warm as exhaust, thanks to the residual kinetic energy of all that roaring air. A low chord rings out from the blackness. A hum.

Now that they have a baseline C_d, Oliver and his team will spend the rest of the day trying to lower the number by making little tweaks to the car's shape with pieces of scrap aluminum and duct tape. If the tweaks work, the team will make them permanent back in Lynchburg. At this point, they can't make major changes to the car—the basic design is frozen, as per X Prize rules—but they can still make small changes. The process might give Edison2 an extra two or three miles per gallon in the competition—perhaps the difference between winning and losing.

Oliver gnaws on his fist as he listens to Fadler make several small suggestions to Barnaby. "These guys are really *smart*," Oliver whispers into my ear. "Shit, man, this is as good as it gets. A $200 million facility!" Oliver glances at Fadler. "A guy like that, this is his dream. He doesn't get to work on cars like this. It's the best car they've ever seen."

They all move to the control room for run number two. Barnaby has made some minor alteration to the radiator; now he'll see if the drag goes up or down. (The radiator is under the hood, but because it alters how air flows through the car, it can affect the C_d.) He calls this process "grabbing increments" and compares it to scraping barnacles off a boat. Remove one or two barnacles and it won't make any difference. Remove fifty barnacles and the boat starts to become lighter, swifter.

The fan spins, the wind roars, the gray console fills with numbers. The Cd is still .171. Barnaby says he's disappointed. "Shows you how extreme what we're trying to do is," he says.

He notices that the right rear wheel pod is vibrating slightly. When the wind stops again, he walks across the test section to a small workshop, where he begins cutting inch-long rectangles out of sheet aluminum with a pair of metal shears. "I'm making vortex generators," he tells me. Moving quickly, lips pursed, Barnaby bends each rectangle 90 degrees into an L and snaps triangles from the edges.

"I knew what was wrong as soon as I saw those wheel pods dance," he says. "I've seen that dance before. It's a very distinctive pattern when you have separated flow. This is where experience comes in." There's aluminum dust on his fingers, and the shoulder of his polo shirt is coated with aluminum dandruff. "You know how they say no battle plan survives first contact with the enemy? No test plan survives first contact with the tunnel. But we're improvising."

Barnaby piles seven pairs of vortex generators onto a silver tray and carries it into the test section, where Ron sits cross-legged, cutting duct tape into tiny strips.

"Look at them," Oliver says of Ron and Barnaby. "They are just like little children, playing with their toys!"

Back inside the control room, I ask one of the GM technicians what he thinks of the Very Light Car. "Pretty low drag," he says with jadedness. "We try to suggest shapes like this to the designers, but the designers say, 'It's ugly.' We say, 'Well, but it can help you.' They say, 'It's ugly.'"

The wind comes on, the wind goes off. Barnaby and Ron decide to make another temporary modification. It has to do with the NACA duct, the inlet that funnels air from the side of the car to the rear-mounted engine and radiator. Air can burble near an inlet, generating turbulence, which results in wasted energy. Some air inevitably spills over the lip of the inlet, like an overflowing glass of milk, instead of entering it. Barnaby now makes an educated guess: Maybe the edge of the inlet is too sharp, and some of the air that's spilling out is getting separated from the car instead of sticking to the body like it should. Separation is bad; separation is turbulence. Barnaby and Brad now use tape to attach a piece of aluminum to the inlet, creating a small curved lip 2 inches wide. By curving the lip, they're hoping to take the air spilling out of the inlet and guide it smoothly onto the body.

The wind blows. A new coefficient of drag flashes onto the gray console: .164, a 4 percent improvement over the baseline.

"Oh, that's *good*," Barnaby says, then refers to himself in the third person: "An excited aerodynamicist in his natural habitat!"

FOR THE rest of the day, as Barnaby continues to work on the car—duct-taping orange yarn to the side to see how it dances, shaping and reshaping the lip on the NACA inlets—Oliver works on Fadler. Oliver is relentless. He keeps up a steady patter of sales talk. He listens with sympathy when Fadler talks about what it's like to work at a company that does good work but that much of the public assumes is a massive failure. By late afternoon, Fadler has softened, and the two of them are huddling in the hallway, Fadler gesturing with his hands, Oliver speaking in a low, understated tone that he reserves for matters of great importance.

During a testing lull, I pull Fadler aside and ask him what he thinks of the Very Light Car. Is it something he would want GM to make? He puts up his hands and smiles. "That's not my job," he says. "I'm not a product planner. From an aero standpoint, this is a very, very exciting vehicle . . . a shape similar to other things that we have considered. However, we never built a prototype."

Fadler says he used to work in GM's race-engineering division, designing high-performance cars for the track, before he shifted to aerodynamics and focused on production cars. (Another example of Oliver's stupendous luck: He gets invited to visit a company with two hundred thousand employees, and his shepherd turns out to be, of all things, *a racing person*.) "There are people who think race cars are complicated," Fadler says. "They are. But they only have to go around the track. Production vehicles have to ride, they have to handle, they have to have comfortable seats, good acceleration, good braking, air conditioner works well, the defroster works well, the seats, the smell, the touch of everything needs to be consistent. Those are the kinds of—let's not call them traps—*challenges* that this vehicle has to face." At the same time, Fadler says, the new federal fuel-economy standards—54.5 miles per gallon by 2025—are "really really aggressive, so we're going to have to move quickly to keep up."

He strikes me as a loyal company man with hopes of changing the GM culture from within, perhaps by using the Edison2 test as a bureaucratic weapon. The next time somebody questions whether a super-low-drag car is possible, Fadler can say, *Yes, and I've got the data right here.* "This is a great opportunity to learn from each other," Fadler tells me. "Put another way: Good engineering is good engineering. It's universal."

A little after 10 P.M., Fadler offers to lead Oliver and a few others on a tour of the rest of the tunnel. Starting at the car, they walk down the slope of the corridor, into blackness, as though descending into a cave. Soon they come to one of the four sets of turning vanes. Fadler says, "The tunnel contains the world's longest aluminum extrusions. Trucked in, assembled in place." Past the turning vanes is a giant curtain that looks like a mosquito net. It's incredibly fine, made of thin metal wires, and rigid. Its purpose is to remove turbulence from the air as it screams through the mesh on its way around the loop. The world's largest screen, it was manufactured in Germany, then shipped to the facility in three sections and welded together.

Oliver looks at the mesh and at Fadler, and it seems like he might start crying. To think that America was almost ready to let GM die! To board this place up!

"Ceiling is 2-foot-thick of poured concrete," Fadler says.

And they haven't even gotten to the fan yet.

Oliver squeezes himself around a narrow corner. The walls open into a vast circular chamber. Before him is the propeller of an ocean liner—the fan—attached to a bomb-shaped nacelle, which houses the fan motor. Oliver approaches it as if in a dream. He stands next to the spruce blades, which dwarf him. He grabs one of the blades with two hands and pushes downward. He has to put his full weight into the blade to get it to move a few inches.

"The fan has six laminated spruce blades with balsawood tips," Fadler says. "The tips go damn near Mach 1. . . ."

Oliver isn't paying attention anymore. The world is different now. He has pierced some invisible membrane. To be such a tiny company, yet to be given the keys to a thing so immense. One of the country's engineering jewels. This *place*. He can't even imagine the expense. A temple of a great bygone culture, left by the old gods of striving postwar America. And to bring an offering that *he* created, that *he* dreamed up, and to see it perform on the level that it has—to have it, in some way, surpass all that came before— ratified at the highest level, at the largest scale possible—it's all too much. He doesn't know what to do with his body. He doesn't know where to put all of the joy.

Oliver takes out his cell and calls banker Mark Giles, one of his investors, back in Charlottesville.

"Mark, I just want you to know, I do not need to talk, but I right now am in the greatest place on Earth. I just want you to hear that."

"I hear you, Oliver."

"The *greatest*, Mark."

"I hear you, and I'm happy for you, Oliver."

"That is all! Bye!"

AT 11:28 P.M., back in the control room, Suzy Cody's phone rings.

"Point-one-seven-one," she says into the phone, settling the bet, and hangs up.

"That was Spock," she says. "I knew he wouldn't be able to sleep."

IT'S ALMOST midnight. After hours of additional modifications, Barnaby and the others have managed to reduce the car's Cd from .171 to .160, an improvement of 6 and a half percent. "That's a big number when you're talking about reaching 100 miles per gallon," Barnaby says.

GM and Edison2 decide to end the day with a smoke test, to get some pictures and see how the smoke flows into the inlets and around the wheel pods. Cody, Fadler, Oliver, Barnaby, Ron, Brad, and I enter the test section. Cody goes into a side room and brings out a long metal tube frothing with smoke.

The parching wind sweeps up at us. The smoke curls over the car in a hard-edged white line. It's almost impossible to see any space between the line of smoke and the top of the car, a sign that the aero is good. Ron and Barnaby cross their arms and yell to each other to be heard above the wind noise, pointing to parts of the car. I can't hear them. The wind flattens Oliver's shirt against his chest, outlining his pads of flesh. He tries to fall over in the direction of the wind, but when he leans forward the wind pushes him back upright.

Cody smokes all parts of the car in turn, the white smoke adhering to the body like water. She smokes the wheel pods. She smokes the front of the fuselage. She smokes the back. The smoke bubbles and pools at one spot toward the rear. Barnaby sees the turbulence and grabs at his beard like it might fly away, his loose-fitting khakis snapping like pennants.

It isn't until 12:30 A.M. that Fadler walks the Edison2 members out to the parking lot. Oliver mentions the irony of today's test: General Motors, a company that shunned the X Prize as an irrelevant distraction, might end up affecting it profoundly anyway, by giving engineering guidance to its front runner.

"If we win by 1 MPG," Oliver tells Fadler, "it may be because of you guys."

Fadler laughs. "And we could be really mean and point that out!"

Instead of saying their farewells, the group drives to a bar nearby, the Victory Inn Bar and Grill. They stay for an hour, drinking beers and talking about aerodynamics, racing, and the X Prize in the place where generations of GM engineers have ended their evenings.

22

The Race

utside a Speedway garage, Oliver climbs onto a bicycle. It's a Monday, the first day of Finals. The sky is pale blue. Gray clouds cast shadows on the infield. A hot wind blows through the garages. Along with the three remaining Very Light Cars, Edison2 has decided to bring a couple of "pit bikes" to Michigan. The guys want to be able to move more quickly from the garages to Pit Lane and back. Oliver's bike is one that Brad brought, a "fixie" with one fixed gear.

"How do I pedal it backwards?" Oliver shouts.

"You can't," the mechanic Bobby Mouzayck says.

"No," Oliver cries, grinning. "Fuck you! That is why there is no freewheel. So you can pedal it backwards. I am going to pedal it backwards." He rides off, pedaling forward.

Everyone is giddy from exhaustion. "I'm tired of this car," Bobby tells me. He's in shorts, his hands smeared with black grease. "So tired. I'm going back to college after this." Mechanical engineering or industrial design, he's not sure. "I'm tired of second fiddle. Being the bitch."

The mechanics aren't happy. After the car came back from the wind tunnel at GM, Barnaby asked them to make some changes. He wanted them to add a paper-thin aluminum lip to each of the two NACA ducts, the inlets on both sides of the car. The lip is supposed to minimize air separation, thereby reducing drag. That's how it worked in the wind tunnel. But the mechanics are dubious about the lip. The wind tunnel is not the track. Besides, they haven't had a chance to test this configuration. If it were up to them, they'd go back to the old inlets. In the words of one mechanic, they'd "drive a stake into the heart of Barnaby fucking Wainfan."

THE FINALS will be almost like another Knockout. All the same tests from Knockout will be performed again: the acceleration and braking test, the durability test, the fuel economy runs, the emissions test, and on and on. The difference is that now, the cars have to hit the final targets—instead of 67 MPGe, 100 MPGe; instead of 68 miles of range for an alternative-class car and 134 miles for a mainstream-class car, 100 miles of range, and 200 miles.

But that's not all. At the end of Finals, there will be races. The point is to break any ties. If multiple cars in a class hit all performance targets, the winner will be decided by a race: 50 laps, 100 miles, with a hard speed limit of 70 miles per hour. Any team exceeding the speed limit will have penalty time added to its total time, and the team with the lowest time will win the millions. Because there are three classes, there are supposed to be three races—two for $2.5 million each, one for $5 million.

At the conclusion of the races, the teams will pack their gear and head to the Chrysler Proving Grounds, twenty-five miles northeast of the Speedway, for the next stage of the Prize, Coast Down. Chrysler technicians will conduct a "coast down" test on each car—a standard way of measuring aerodynamic drag and roll-ing resistance that involves bringing a car up to 80 miles per hour then letting it coast down in neutral. Then, in August, for the last stage of the Prize—called "Validation"—the teams will travel 240 miles west, to the U.S. government's Argonne National Labora-tories in Lemont, Illinois. Argonne is an American landmark, an outgrowth of the World War II–era Manhattan Project. Its scien-tists designed some of the world's first peacetime nuclear-power reactors. Today Argonne is home to a lab that researches advanced vehicles. There, the cars will be strapped to dynamometers and driven in controlled conditions, their performance measured by machines.

Finally, in September, the Foundation will announce the win-ners of the Prize at a ceremony in Washington, D.C. Officials have decided not to fuss over the winners here at the Speedway. They're saving the party for later, when they can put on a show for the nation's politicians.

But the teams know that this week is the crucial one. It's the last stage on the track, the last chance to prove the cars' mettle in real-world conditions. This is why there are several newcomers in the garages today. Certain family members and friends have made the trip to Michigan for the first time. One is Kim Kuttner, Oli-ver's wife.

"Hi, schnooks," she says to Oliver. "How are you?"

Oliver has just slumped into a folding chair on the lip of the garage. He's looking glumly out into the sun. Kim is wearing dark round sunglasses, skinny jeans with no pockets, and gold shoes.

"Okay," Oliver mumbles.

"Are you nervous?" Kim asks.

He shrugs. He starts telling her a story. He just cut a deal, he says. He made a gentleman's agreement with one of his opponents, Roger Riedener, makers of the Swiss E-Tracer motorcycle. Oliver and Roger are competing against each other in the $2.5 million alternative-tandem class. And next Tuesday, they will race: Roger's two yellow motorcycles against Oliver's Very Light Car.

The deal they've struck is this: If Oliver wins the race, and the $2.5 million, he'll give $500,000 to the second-place finisher, $300,000 to third, and $200,000 to fourth. If Roger wins, he'll do the same. "Basically, I want to bet against myself," Oliver says.

Kim shoots him a distant, admiring look. "You're such a *businessman*," she says.

"Well, I run the numbers."

Oliver tells Kim he sees the deal as an insurance policy. He feels he needs one, despite the fact that, on paper, he seems to be in an enviable position.

At this point, his two mainstream cars are the only ones left in the mainstream class. His last competitor was Kevin Smith of Illuminati, and Kevin failed the acceleration test at Knockout. There will be no race, then, to determine the winner of the $5 million mainstream class. Why hold a race when Oliver would only be racing against himself?

Now he has to do only two things to win the $5 million. One, he has to hit the performance targets on at least one of the cars, both here at the Finals and at the stages to follow. Two, he has to get that car through without breaking it, which is probably the harder problem, because each new test is another chance for a Very Light Car to blow a piston or a tire, for its bluebird of an engine to decide it has taken quite enough abuse already and would rather expire.

Each car is a scalpel. How many cuts until it grows dull?

Hence the insurance policy. If Oliver's mainstream cars break and he loses the $5 million, he could still win the $2.5 million tandem prize. And now, thanks to his deal with Roger, even if he loses the tandem race, he won't go away with nothing.

He *can't* go away with nothing.

It's been a difficult few months for Oliver's real-estate business.

The banks aren't lending. He has pared his operation back to the bone. "I don't hire anymore," he tells Kim. He used to employ many building repair crews. Now he has just two.

"And yet when the side of the Linen Building was falling down over the winter," Kim says, "you went over and repaired that."

Oliver nods. His face is stone. "Ten guys," he says. "I have ten guys now, other than this. That's all I have. . . . It's my sustainable construction company without banks. Anything bigger is not sustainable."

PAUL WILBUR, CEO of Aptera Motors, comes wandering through the garage. Aptera's car is still eligible for one of the $2.5 million prize pots. Paul is the former Detroit executive who took over the company from its founder. Many of the earliest fans of Aptera dislike him intensely. I've never spoken to Paul, so I introduce myself.

The first thing he tells me is that even though his company is based in California, he still lives around here, outside of Detroit, with his wife and his youngest kid; the other four are in college.

The second thing he tells me is that his house is worth a million dollars plus and he can't get rid of it in this market.

Paul says he admires all the people here. They're all innovators, just at different stages of the process of making something real. He mentions John DeLorean, the former Detroit engineer who started his own company, and says I should read his biography.

"Didn't DeLorean fail?" I ask.

Paul looks surprised. DeLorean built a factory; DeLorean manufactured cars. What sunk him was something more prosaic: "DeLorean ran out of money."

Paul looks at his watch. He says he has a phone call with the Department of Energy. Aptera's been trying to secure a government loan for two years now. "We're in due diligence with them," he says, and heads back to the Aptera trailer.

A little while later, I walk out behind the garages and see a small trailer I never noticed before. The inside is lined with bikini calendars. It belongs to the team from Li-Ion Motors, one of the two pariah companies in the field. Li-Ion used to be called EV Innovations; before that, it was Hybrid Technologies. At one point it also sold Internet phone service. The founder of Li-Ion lives in Canada, and the corporate office is in Las Vegas, but the guys in the trailer tell me they're from Moorestown, North Carolina—NASCAR country. Their driver has driven for NASCAR teams. Their battery engineer graduated from North Carolina State. Their lead mechanic used to race dragsters, and their team leader used to make cars for the movies, including some of the cars in *Gone in Sixty Seconds*. They say they don't have any idea what's happening at corporate.

During Knockout, their car, the Wave II, shot to the top of its class, quietly achieving 182.3 MPGe. Unlike the Aptera 2e, which was designed from scratch, the Wave II is a mash-up of the original and the off-the-shelf; the aerodynamic body is custom, but other parts of the car come from BMW, Smart, and Honda.

"Found the sweet spot," one of the guys tells me. "It's not a lot of trick stuff. If something went wrong, it wouldn't be a crazy price to fix."

OLIVER GETS the news about the tandem car on the third day of Finals. As soon as he hears, he rushes to tell Bobby. It's only right. It's his car.

"Your car got 130 miles per gallon on the highway cycle," Oliver says.

Bobby laughs.

"A hundred and thirty fucking miles per gallon," he says. "Was that running the EGR at all?"

The EGR is the system that recirculates a portion of the

exhaust gas back through the engine, improving fuel economy; because it adds a layer of complexity, the team sometimes leaves it off, to boost reliability.

"No," Ron Mathis says.

"Perfect," Bobby says.

Oliver says, "No one can say we can't do 100 miles per gallon on gasoline. We know we can. We are there."

"A hundred and thirty fucking miles per gallon!" Bobby says.

"Those are the numbers," Ron says.

OLIVER'S MOTHER and stepfather, Beatrix and Ludwig, arrive on the morning of the fourth day of finals. He's been stressing about this for days. He loves them, but they're his parents. Having them around means more to worry about, more to manage.

"Olly, please, a picture, a picture by the car."

Behind one of the garages, Beatrix holds up a camera, trying to get her son's attention. Oliver obliges. The two of them pose in front of a Very Light Car, Oliver in his white Edison2 shirt, his mother in a billowing silk shawl, the reds of which are set off by her purple hair.

Ludwig looks on, wearing a tan pinstripe suit, a pinkish striped shirt, and glasses. He flashes a charming smile; his teeth look like they could crack marbles. "It is like I always say," Ludwig says. "Centuries ago there was no sanitation. They would hang a little bag out the window and shit. They said it was stinking, it was terrible, but they did not do anything. And then the plagues came, so, 90 percent death in some towns. And they said, *now* we must do something about it."

I ask him what he thinks of Oliver's X Prize bid. He must be very proud. "He is like an artist in some ways," Ludwig says. "It is what our family does. You never say, Oh, *now* I made it. You *never* made it. It's a constant destruction."

THE CARS are all together for one of the last times. Laps and laps. Some split the air as silent as doves, others sizzle like eggs, their hoods throwing slanted light onto the perimeter wall. High above the stands, a few stranded birds tip their wings and soar over the infield toward the charging tent.

Brad Jaeger, driving one of the Very Light Cars, slows to a stop; this is a highway run, which requires periodic stops and starts. He catches up to fellow driver Emanuele Pirro at the line. Brad points his smartphone at Emanuele and takes a video. They drive side by side for a time; then Brad lets him back in front. When Brad watches the video later, he can see the reflection of his own car slide liquidly across the chrome-colored panels of Emanuele's.

Later, there's a final range test. After 100 miles, the alternative-class cars pull into the grass, one by one, to jubilant cries from their makers. The Very Light Cars have to stay on the track until mile 200. Emanuele finishes first, followed by Brad. The men of Edison2 crowd together in Pit Lane to greet the drivers. Emanuele flashes a peace sign to a video camera, spritzes his sweat-slick hair with a water bottle, and hands a piece of paper to Oliver. There are three neat columns of tiny numbers written in blue ink. During the event, he leaned down and recorded his own lap times on a piece of paper strapped to his right leg. Most of the lap times are within a second or two of each other.

"*That* is a professional," Oliver says, marveling at the neatness of the handwriting.

"It's to keep track, you know," Emanuele says, humbly.

There's some hugging, some slapping of backs.

"Ah, you are the best!" Oliver says. "Best team!"

THEY HEAR numbers from the officials. The numbers are good. They hear 100 MPGe, they hear 95, they hear 109. They seem to have hit the 100 MPGe target on the tandem car and at least one of the mainstream cars. Which means they've done it. They've won the $5 million—maybe. The ultimate decision is still in the hands of the officials. And after this, there are still other stages: Coast Down, then Validation. Always another stage, another test. This is why no one on Edison2 is celebrating: The Prize has conditioned them to be wary of anything that feels like finality. It's been the engineering equivalent of a presidential campaign, with all of a campaign's grueling traits, and with the final decision still months away.

"It's all so fucking anticlimactic!" Oliver says.

But at least he still has a race.

The tandem race. Emanuele driving Bobby's car. Emanuele versus the two Swiss motorcycles for $2.5 million.

That weekend, the team plots strategy in their hotel.

On Monday morning, the day before the race, Emanuele climbs into Bobby's car to complete one last test: 0-to-60 acceleration. Earlier, Ron, Brad, and James Nero were testing a feature of the car called "launch control." The point of the launch control is to help the car get off the line quickly without damaging its tiny one-cylinder engine. It's basically a fail-safe system, programmed into the car: You sit there with the car stopped, in first gear, with the clutch depressed and the throttle open all the way, and once you release the clutch slightly, the car goes through a sort of launch sequence, revving the engine in steady increments without letting it "overrev."

It turns out, though, that there's an error in the program—and now the engine overrevs, zooming from 7,000 RPMs straight to 15,000. Emanuele tries to come off the line and hears the sickening crunch of an engine over-revving itself to death.

When I talk to Bobby about it later, he doesn't seem that upset about the destruction of his car. "We went out fighting," he tells me. "That's the important part. It wasn't that we lost a wheel, or didn't put enough oil in." He says he's proud of the tandem car's achievements, especially its city-cycle MPGe: 92. "That's fuckin' real," he says. "That shit really did it. And you put some reliability into it and whatnot, this car will fuckin' do it, full-time. I mean, if you could drive around New York City ninety miles on one gallon of gas, you'd be the wealthiest cabdriver in town."

The over-revving incident puts an end to any sort of Edison2/X-Tracer confrontation. There will be no race between the Very Light Car and the Swiss motorcycles. The Swiss motorcycles are now the last vehicles left in the tandem class. Assuming they make it through Validation, they've won their class, and the $2.5 million. The gentleman's agreement between Oliver and Roger is also canceled, since Oliver no longer has a car to race. This means that the most money Oliver can take home now is $5 million. It also means that of the three races that were scheduled for the end of Finals, only one will actually take place—a race between the five remaining cars in the alternative side-by-side class, all of them electrics.

Around lunchtime on Tuesday, the day he was supposed to race, Oliver gathers his team in a circle behind the garage. He wants to thank them. Edison2's job here is done. No more tests, no more runs. They're pretty sure they got the numbers. All that's left is to watch today's race between the electric cars, then pack up the gear.

"I don't know what to say," Oliver says. "I love you guys. You did an awesome job." He tells them the race might be interesting. On the far straightaway, the X Prize has set up a chicane—a tight series of turns. The chicane is marked by cones. Through the chicane, the electric cars will have to brake and accelerate, brake and accelerate.

"My personal gut feeling is somebody's gonna get hurt," Oliver says. "But that's just me. The chicane is where it's gonna happen if it happens. And I hope it doesn't."

A few hours later, as a direct July sun heats the asphalt, Oliver and his crew gather near the starting line. The five electric cars are lined up. The race is about to begin.

Officials release the cars at ten-second intervals: first TW4XP, then a car built by a Finnish team called RaceAbout, then Zap, then Li-Ion, then Aptera. There's a smattering of applause. Then the spectators start walking across the infield to the chicane. Oliver gets there in time to watch each car screech through the cones on the third lap. RaceAbout's car—created by university students and professors from Helsinki—seems to be fastest. It's a burly, heavy coupe with not one but four electric motors, one for each wheel. RaceAbout is followed by Li-Ion and, surprisingly, TW4XP. The little German car is nimble.

Oliver says, "You can hit these turns pretty hard, but not too many times, because you're draining down your battery."

His stepfather lights a cigarette. His mother sits on a golf cart wearing an enormous straw hat.

"The real question is, are they going to finish?"

Oliver isn't talking to anyone in particular. He looks around. He is standing in a muddy ditch a few yards from a Porta Potty. There are about twenty-five people here.

"This is no different than in 1910 and the steam engine takes on the electric!" he cries. "It's crazy, it's crazy!" Oliver cries. "Two-point-five million. There is none bigger. It is the race of the century."

Already, Li-Ion has moved up to second place, passing everyone but RaceAbout. Aptera has pitted. Aptera's support team darts to the car to see what's wrong.

A man with an iPhone is timing the laps. He calls out that the gap between RaceAbout and Li-Ion is 13 seconds.

Oliver says, "This is going to come down to no time at all."

With nineteen laps to go, Oliver gets a call on his cell. It's a reporter from the *New York Times*. "Our car is made of steel and aluminum," he says, and reels off what by now has become his

catchphrase: "The car is light because it is light." Then he begins
to give a play-by-play of the race to the *Times* guy. "This is a race
for $2.5 million, and no one is watching. What is the differential?
Two seconds? Two seconds, for $2 million. It is a fair race, a well-
designed race, between a Finnish team and a team from North
Carolina."

Oliver scribbles lap times on a hotel pad.

"No wind today! That is good. This place is a vortex. You can
get nailed. We could have gone 68 miles per hour. . . . They are
going 66 miles per hour. We would have gone 68. We would have
risked more. . . ."

The gap is 2.5 seconds now, with two laps to go. Zap is in the
pit, its battery pack depleted. It won't finish the race.

Last lap. The guy with the watch screams, "One second, one
second, one second."

Oliver turns to Brad: "One second!"

On the other side of the track, blocked from view, the cars
cross the finish line within .0179 seconds of each other. A judge's
walkie-talkie crackles. "RaceAbout," she says, calling it for the
Finns. Later, though, the judges will determine that the Finns
have broken the speed limit twice. The penalty time changes the
outcome of the race. The jackpot will go not to the Finns but to
Li-Ion Motors, the penny-stock good ol' boys.

Oliver will call me a few days from now to make sure I understand what I saw. "It was the race of the century," he'll say. "I could
feel the tension in the drivers' meeting in the morning. It was one
of the most beautiful stories ever."

The traffic cones and the mud. The faint reek of the Porta
Potty.

We walk back across the infield to the garages.

23

Capitol Hill

Four days after the race, I get a call from Oliver. He's on the road with Ron, driving from Michigan back to Charlottesville. Oliver says he has something important to tell me. Important and hilarious. "You're gonna die laughing."

The cars broke, he says. Both of the remaining Very Light Cars. Dead. Pistons blown, engines destroyed. They broke at the Coast Down stage, at Chrysler's test facility, when the Chrysler technicians tried to drive them.

For weeks, actually, Oliver has been telling me this might happen, but I haven't paid much attention. He made it clear he hated the idea of people other than Brad Jaeger and Emanuele Pirro driving his cars at Coast Down. The Prize was refusing to let him use his own drivers, and he was livid. This seemed natural to me. What creative person doesn't freak out a little when handing his creation to a stranger for the first time? But now, Oliver says, his worst fears have been realized. "To us, these cars, they're like having made a painting. Somebody just destroyed, really, an engineering piece of art."

It's hard for me to make out exactly what happened; Oliver's talking too fast, cursing too much, drawing line after line in the dust: "I want it in writing, in writing, that you"—the Foundation—"take full responsibility for it. . . . I want them to be able to dig themselves out of their own stupid shit." He passes the phone to Ron, who is serene as always. Ron explains that the Chrysler guys "were just a delight" to work with; they simply made a crucial mistake. To shift up a gear in the Very Light Car, you pull the shifter toward you; this is the way it works in race cars and some street cars. But many street cars, including Chryslers, are the opposite; to shift up, you push the shifter away from you. The Chrysler drivers moved the shifter in the wrong direction at the wrong time, due to muscle memory. They downshifted instead of upshifted. Imagine riding down a big hill on a bicycle and suddenly shifting into first gear. Now imagine that your legs are locked into the pedals and *forced* to pedal at first-gear speed. Your legs would break. The bike would break. When the drivers downshifted, the engines of the Very Light Cars over-revved and tore apart. Now there's no time to manufacture the custom engine parts that Edison2 would need to repair them. Ron says the responsibility is now on the X Prize to "make a decision, for the good, that will satisfy all parties." He adds, "The threat of a lawsuit is hanging like a sword over all this. . . . Now the X Prize has gotta step up and say, 'Well, blimey, nobody expected this.'"

Ron returns the phone to Oliver, who tells me what will happen next. The remaining cars will go to Argonne National Laboratories next week, to complete the Validation stage on Argonne's dynos. But Edison2 obviously can't put its cars on the dynos, because the cars are dead.

In lieu of the test, then, Edison2 will submit a report to the X Prize. All along, the team has been testing the Very Light Cars at another lab, Roush Laboratories, and Oliver thinks he can use the Roush data to show that the cars have *already* been validated.

Far from being humbled by the destruction of his cars, Oliver seems indignant about the extra work he must now perform to show their worth. "It's like a kid who wants to play soccer, and he's told he has to do more homework," he says. "In this case, the teacher *burned* the homework. We've got a photocopy, we give the photocopy. Take the photocopy and let me go play soccer. We want to play soccer. It's for the good of the world." Cooling down, he adds that he's grateful to Diamandis and the Prize for putting him "in an interesting, enviable spot. You know, this is the rocket that blows up. When Apollo started putting rockets in space, a few of them blew up."

I'm not able to talk to the Chrysler drivers to get their side of the story. But it seems self-serving for Oliver and Ron to pin the blame on them. One of the rules of the Prize is that someone new to the car has to be able to figure out how to drive it. And the Chrysler guys are hardly amateurs. They do this for a living.

I do eventually talk to someone who says he spoke with the Chrysler drivers: BITW Technologies' George Voll, the Indiana hardware-store owner. George's diesel Chevy Metro was eliminated during Knockout for falling well short of the fuel economy target. "Oh, yeah, they were pissed," George says of the Chrysler guys. "They drove it just like Oliver told 'em to." Ultimately, though, George doesn't care *why* the Very Light Cars blew up. To him, the crucial fact is that they did. They went over the ragged edge. "Where I come from, if you fail, you lose," George tells me. "And Oliver failed. He was marginal at best with his technology."

He adds, "They're gonna say that I'm a sore loser. And I'm sore. Yeah, I'm sore about it. But I see him taking people's money. That car could not go two hundred miles without blowing that engine. Guaranteed.... And he was such a goddamn crybaby through the whole thing! Reminded me of a child.

"He had a lot of money. My father would tell me that the guy with the most money's gonna win."

OLIVER DOESN'T hear anything from X Prize officials until the week before the winner's ceremony, when he calls them to ask what's up. They tell him he "should be good" but don't say much more. He's pretty sure the Foundation is going to give him the $5 million, but not entirely.

On September 15, the eve of the winner's ceremony, he and Kim arrive in Washington, D.C. They check into a nice hotel that the X Prize has arranged for them.

His parents, Ludwig and Beatrix, are also in town. They've come to see Oliver accept his prize.

They all go out to dinner at a burger joint. Kim and Ludwig fall into an intense argument of some kind. Oliver tries to ignore it. He starts to jot down notes for his acceptance speech on a spare piece of paper.

After dinner, he and Kim go back to the hotel and go to sleep.

In the morning, she tells Oliver she's decided to leave him. She wants a divorce.

Oliver doesn't raise his voice. He doesn't cry. He's surprised, but he wonders if this is real. Maybe it will blow over in a couple of weeks.

He tries to ignore it. He shoves it down. He might have to give a speech today to congresspeople. Kim is still coming to the ceremony.

He focuses on the speech.

A little before noon, Oliver and Kim make their way to the Historical Society of Washington, D.C.; its building, a stately Beaux-Arts structure, used to be known as the Carnegie Library, its construction funded by the famous steel magnate. Prize officials have set up rows of white folding chairs on the lawn, facing a Lucite podium and a sky-blue X Prize banner. Oliver wears black jeans and a white Edison2 shirt. Kim sits at his side, looking radiant in her own Edison2 shirt, skinny black pants, and dark sunglasses. They sit there and smile like nothing has happened. (Later, when I ask Kim about the events of that morning, she writes in an e-mail that she can't go into details "without hurting many people involved," adding, "I believe that in families and in relationships there are always many sides and many different views. That being said, I feel that in order to get to the point of a divorce, a long history of problems must exist between the two parties.")

The other winning teams are here, too—X-Tracer, from Switzerland, and Li-Ion Motors—but several additional teams have also made the trip, hoping to chat up the powerful and maybe give a congressman or two a ride in their vehicles. Jen and Kevin from Illuminati are here with their car. "Wouldn't have missed the fun," Jen says, grinning. Same with Aptera and West Philly. Aptera hauled the 2e from California and parked it next to the lawn. West Philly didn't bring either of their cars, but a handful of students and adults are here to represent the team. Toward the back of the lawn is a line of TV cameras on tripods. I spot the CEO of Progressive Insurance, walking around with an X Prize pin on his lapel.

At noon, Diamandis appears behind the podium in a white shirt and paisley tie and kicks off the event with his standard paean to "visionaries and doers," his voice booming through the PA: "We live in a time and age when anything is possible. Where a man or a woman who is passionate or driven can go out and build a spaceship, or a 100-mile-per-gallon-equivalent car, or reinvent how we educate or how we feed the hungry. . . ."

He yields to a series of politicians. Democratic Speaker of the House Nancy Pelosi points out the central role of science in President Obama's agenda: "Four words . . . science, science, science, science." Ed Markey of Massachusetts, a wiry, sixty-four-year-old Democrat fervent about energy independence, pumps his fist and uncorks a stump-speech zinger about how America needs to stop being "OPEC's ATM." "Our strength is our ability to invent," Markey says, telling the teams that they are the Lindberghs of the modern age. The actual award presentations don't begin until Robert Weiss, vice chairman and president of the X Prize Foundation, takes the stage. Weiss directs the audience's attention to a large video screen. A soaring tune plays, and a number flashes on the screen in neon-green digits: 205.3 MPGe, the final fuel efficiency of the Swiss motorcycle. He calls the Swiss to the stage and presents the team's head of marketing with an oversized check for $2.5 million. "What we cannot do is mass-produce our vehicle," the head of marketing says. "We are just too small a company. . . . We know a lot and we are willing to share the knowledge. So please come talk to us."

Weiss repeats the process for the second alternative-class winner, Li-Ion Motors: another video reveal of the final MPGe (187 MPGe), another $2.5 million check. Then he starts talking about the Very Light Car.

The Very Light Car, he says, "has proven itself highly efficient and safe." Its final efficiency appears on the screen: 102.5 MPGe. The Prize, it turns out, has accepted Oliver's argument that his cars broke at Coast Down due to driver error, and only after hitting the Prize's targets on the track and in the lab. Oliver had made his case to officials in a report full of numbers from Roush Laboratories and the cars' own data loggers. The data were persuasive: amazingly, both of his four-seat cars achieved 100 MPGe—100.2 MPGe for one, 102.5 MPGe for the other—and emitted far less than 200 grams of carbon dioxide per mile, despite running on internal combustion engines. Oliver didn't just squeak through; he cleared all the hurdles with a bit of room to spare.

Which means that Edison2 has won the $5 million. Far more important, though, Oliver and his team have proved that a completely new kind of car is possible. With a couple million dollars, they have taken apart the automobile and put it back together in a lighter, fleeter form—a feat that has eluded almost every major automaker, even with all their billions. The quest required talent, vision, risk, and luck: Barnaby's eyeball aero; Ron's perfectionism; the physical endurance and improvisations of the mechanics; Oliver's tenacity; the GM wind tunnel test. And all of that crazy-ass energy and effort was summoned from nothing, three years ago, by a guy using a huge cash prize as bait. The contest worked. Diamandis went looking for an innovator who could change the world if only the world would let him, and he found Oliver Kuttner.

To rousing applause, Oliver ambles to the podium followed by Kim, twelve employees of Edison2, and Senator Mark Warner. An X Prize employee emerges from the side of the stage holding the winner's trophy for the mainstream class, which resembles a gyroscope eating a globe. Someone else holds a ceremonial check for $5 million, signed by Diamandis.

Onstage there's a lot of clapping and slapping of backs. The senator gives a brief speech. "We need a few less financial engineers and a few more real engineers like these people back here," he says. "These are the next great innovators that will lead our country forward." He turns to hug Oliver, who then makes his way to the podium.

Oliver sets his notes down. He looks out at the crowd of 150 gathered on the lawn.

"I am going to deviate from the script for one second," he says.

Oliver could lead with anything. Lightness. Aerodynamics. The struggles of an entrepreneur on the ragged edge. But instead he talks about the importance of Lynchburg to the Very Light Car. Given a chance to address some of the most powerful politicians in the country, he wants to talk about machine shops. He says he built the car in Lynchburg because it was where he could get custom parts made quickly. "You must be able to go to someone to realize the idea," Oliver says. "As a nation, we face the risk of losing some of those people who can realize the idea. It is imperative that we change the course of this nation and concentrate on real engineering. . . . We need to get over partisanism. The world needs to push for innovation."

Then he pauses and picks up his notes. He mumbles at first, head down. He thanks his team. He says he's so proud of them. Now he looks up.

"We propose a rescaling of the automobile. We propose a lightweighting in an unprecedented fashion." He and his team know how to make the rescaling work, Oliver says, and this knowledge is vital to the industry. There are people in the industry "who don't know, who are going to spend billions of dollars and may call us in tears," he says. "We don't take pleasure from it. But facts are facts." His employees are grinning now, basking in Oliver's bombast. They earned it, this moment. Oliver gains confidence and speed as he brings the speech to a close. People will come to believe in Edison2's philosophy, he promises. "It will be a minority initially, and then it will be a storm."

Making his way off the stage, Oliver is swarmed by video cameras, like a quarterback in the locker room. I hear him say, "The car is light because it's light."

"What are you going to do with the money?" a reporter asks.

"We are going to expand," he says. "We are looking at *large* facilities."

Nearby I notice Mark Giles, one of Oliver's most loyal investors. I ask what he thought of Oliver's speech. "He was a little off today," Giles says, "but he's also damn tired." Giles says the investors are ready to put in another $3 million. "We want to see a balance sheet and some staying power," Giles says. "We want to get some of these babies driving around. . . . We're working on this path to profitability. And if we don't make progress, I believe that Oliver is perfectly capable of pulling the plug."

———

AFTER THE ceremony, a few of the West Philly kids and adults head over to the Eisenhower Executive Office Building, next door to the White House, where they've been invited to listen to President Obama deliver a speech about science education. From the podium, the President gives the team a surprise shout-out, praising them for getting so far in the X Prize. "Now, they didn't win the competition," Obama says. "You know. They're kids. Come on." Still, West Philly went "toe to toe with car companies and big-name universities, well-funded rivals," and "held their own."

Meanwhile, Oliver and the other two winning teams make their way to Capitol Hill. The Foundation has pulled some strings to help them make their case to Congress, setting up a hearing before the House Select Committee on Energy Independence and Global Warming, chaired by Ed Markey. In the Cannon Office Building, in a high-ceilinged room hung with six ornate chandeliers, the three team leaders and Peter Diamandis sit at a long brown table, facing a navy curtain on the other side of the room. An audience of forty files into rows of seats behind them. Markey enters the room and takes up a perch in front of the curtain, and other Congressmen soon follow, along with their aides. One is Tom Perriello, a young Virginia Democrat who recently visited the Edison2 shop in Lynchburg.

This is "a very historic hearing," Markey tells the panelists, speaking into a microphone. "Consumers are eager to hear how soon these vehicles can get from your garage to their driveways." Diamandis leads off, introducing the winners. "These are not experts," he says. "An expert is somebody who can tell you exactly how something *can't* be done."

"Your Honor," Oliver says, addressing Markey, "we are perhaps somewhat patriotic. . . . I'm a German national. I want to see the jobs come back. I want to see the factories full." Oliver keeps his remarks high-minded, sticking to broad themes; Markey leans forward whenever he hits his points. After a time, Oliver yields to the other team leaders, who spend much of their presentations asking for money. Markey glazes over.

Walking out of the hearing room, I run into Kim. "He *stuck it*," she says, forming two fists and pumping them ecstatically. "I'm a perfectionist. I want things to be *perfect*. It's gymnastics. He stuck it."

Kim and I step out onto New Jersey Avenue, where the three winning cars are parked side by side. Markey soon emerges with two aides and climbs into the driver's seat of every car while the aides take pictures. It's a delicate seven-man operation to get the elderly Markey in and out of the tiny E-Tracer. "Well done," he says to no one in particular. "We want to help you guys." Then he waves, and he and his aides walk away.

Someone hands Oliver a phone. He starts talking excitedly about the next iteration of the Very Light Car. Oliver calls it, somewhat confusingly, the 4.0. (The 1.0 and 2.0 were pre–X Prize prototypes, and the 3.0 was the X Prize car.)

"The next car is going to become a car that you will recognize as a daily car," Oliver says. "We know we can do it."

Giles says to Kim, "Is he on with the *Wall Street Journal*?"

"The *Wall Street Journal* or the *New York Times*."

"Good job," says Giles, looking at his watch. "Chu. Chu is next. Let's go to Chu."

Steven Chu: Obama's secretary of energy. Promoter of electric cars and alternative energy. Booster of ARPA-E, a new government program to support radical advances in science and engineering. Co-recipient of the 1997 Nobel Prize in Physics for inventing an ingenious scientific tool. A young Chu figured out how to use multiple lasers to cool atoms so that they slowed way down, moving in a kind of "optical molasses"; individual atoms could then be trapped and studied in great detail.

Chu's agency has supported the auto prize all along, contributing $5.5 million in federal stimulus funds and a $3.5 million grant for education and outreach. No one knows how carefully Chu himself has been following the competition. But Perriello has done a favor for Oliver by asking Chu to take a quick look at the winning X Prize cars, and Chu has apparently agreed. It's a huge opportunity for all three of the teams; a kind word from Chu could release millions in loan guarantees or other government funds.

The teams drive their cars in a Dr. Seuss procession to the headquarters of the Department of Energy a mile west, turning the heads of other drivers en route. They park in a line outside the main entrance. A DOE aide in a suit and a purple silk tie emerges from the building and tells them that Secretary Chu will be out shortly.

Fifteen minutes later, there's a small commotion as Chu makes his way to the cars, flanked by two aides. He is short, thin, and doll-delicate in a dark suit and yellow tie. Oliver intercepts him.

"We have totally reengineered the way that a car is built," Oliver begins. "No one has fully understood this."

Chu, the scientist, recoils from these grandiose words. He leans back, sticks his hands into his pockets.

"Are you familiar with sports-car racing?" Oliver asks.

Wouldn't you know, the Nobel Laureate is not familiar with sports-car racing.

"Are you familiar with races like the 24 Hours of Le Mans?"

Chu shoots Oliver a blank stare.

They aren't so different, actually. As younger men, they were both rejected by Ivy League colleges. Chu went on to work at Bell Labs, where he was known as a hands-on experimentalist instead of a strict theorist. Oliver has always dreamed of working at a place like Bell Labs—a playground for adults. Both are hackers at heart. But they can't communicate. Chu now reaches down and gives the wheel pod of the Very Light Car a shove. Then he shoots Oliver a puzzled look, as if to say, *What is this? Why is the wheel like this?* Oliver responds by lifting his right leg and pressing down on the same wheel pod with his foot, as if to say, *What do you think you are doing to my wheel?*

Then, with a curt nod and a barely audible "thank you," Chu walks away from the Very Light Car to check out the Swiss motorcycle. The entire encounter hasn't lasted longer than a minute and a half.

"Doesn't get it," Oliver says, turning back to Kim and Ron, his face frozen in a pained grin. "I thought he was smarter than that."

Oliver looks out toward Independence Avenue. For a moment, he watches the cars come and go. Sedans, pickups, SUVs. A gray waste. So much heedless metal, so many rough holes punched in the air. When he spins back around, he sees Chu talking to the marketing manager for the Swiss motorcycle. Chu is smiling. His lips are making the shape for "wonderful."

After a long pause, Oliver says, "At least we got good pictures."

Kim looks at Oliver with sympathy. "Nobel," she says, almost spitting it, trying to make Oliver feel better about the eminent man's lack of perceptiveness. She puts a hand on his shoulder.

On the far side of the Very Light Car, a businesswoman in black pumps and a black jacket slows as she walks by the car, running her eyes along the fuselage. Oliver lunges toward the woman, intercepting her with a big, sloppy, friendly smile.

"This is eight hundred pounds," Oliver says. "Push it."

24

Traction

Twenty-six days later, on October 12, Oliver is in Lynchburg, behind the wheel of his Jetta, approaching one of his many properties in the city. He has an appointment with someone who might want to buy the property. He's running late when a police officer pulls him over.

Oliver gets out of the Jetta and walks toward the officer. "What did I do?" Oliver asks. The officer tells him his registration tags are expired, and he should get back in his car.

He's been having a tough few weeks. His split with Kim has turned out to be real and irreversible. There's another man in the picture now. Kim met him in Prague, in the Czech Republic. He's apparently a real-estate developer, like Oliver—but younger, in his thirties. A dark-haired, handsome Italian.

Now this.

"Give me a break," Oliver tells the officer, walking back to his car. "I'm trying to fix your fucking city."

After sitting in the car for a minute, Oliver opens the door and just starts to walk away, in the opposite direction of the officer, toward his appointment. The officer tells him to come back. Oliver complies. But now other officers start to arrive—two, then three, then four. They see a large man muttering in a German accent, arms folded, defiant. They take flanking positions. Oliver starts to yell and point his finger. He shouts, "Get the fucking news down here. . . . I want them to see how stupid these people can be." Typically for Oliver, he also apologizes to the officers, telling them he's been having a bad day. But he refuses to get back in the car. The officers say they'll arrest him if he doesn't.

"Get off my property," he says.

At this point they handcuff his arms behind his back. Two officers drag him to the back of the cruiser. He cries out that they're hurting his shoulder.

The arrest makes news in Lynchburg and also back in Charlottesville. X-PRIZE WINNER ARRESTED IN LYNCHBURG, reads one headline. Convinced he should have gotten away with just a ticket, Oliver complains about the arrest to Lynchburg's city manager. The city launches a probe. But the whole incident was captured on the police cruiser's in-dash camera, and when the police release the footage, it's clear that they acted appropriately. Authorities charge Oliver with disorderly conduct, a misdemeanor. He pleads guilty, paying a $250 fine, and in court he apologizes for being "out of line." In the comment threads of newspaper stories about the arrest, people are split on whether to blame Oliver for his arrogance or the Lynchburg police for their zealousness.

I see Oliver not long after the arrest, when he's driving through Philadelphia with David Brown. The three of us meet for lunch at a bistro downtown.

Oliver orders an Italian sandwich and leads off with some surprising news: He has bought a factory. One hundred fifty-seven acres on the James River in Lynchburg. It's an old steel foundry. He bought it on a whim, without even knowing if there were any toxic chemicals in the soil. He simply drove past the site, stopped his truck to investigate, and decided he had to have it.

"You *have* to see it," Oliver says, leaning across the table, eyes shining. "Basically, I bought a ruin. I bought post-apocalypse. You have to see it and feel it, because two years from now, it will be a jewel."

He plans to renovate the factory into a new headquarters for Edison2, complete with a test track and extensive R&D facilities. Then, almost in the next breath, Oliver says his new goal is to "be more mobile. My properties own me right now. I don't own them." He takes a large bite of his sandwich.

He says he's been thinking about whether he would still participate in the X Prize if he had it to do over again. He's not sure he would. "It pushed my wife over the edge," he says. He isn't sure what will happen now with their kids. The other day, he saw a family of four sitting at a coffee shop. "And I just watched them," Oliver says. "They didn't say much. But they were a nice family. They were kind of . . . together."

He waves his arm. His voice quavers.

"And I was thinking I'm never gonna do that."

He starts to cry. Noisily, unself-consciously. His cheeks are red and puffy. People in the bistro turn and stare. David Brown looks at his lap.

After a few moments, Oliver dries his face with his napkin. "It is what it is," he says, frowning. "I hit one target and missed another."

Now Oliver shifts the conversation to Edison2. He wants to talk about the future of his company. But I notice a difference in emphasis and tone. Until now, all of his appeals to potential investors have been patriotic, along the lines of his speech at the winner's ceremony: If we want dramatically better cars in America, we have to invest in American companies, American machine shops, American manufacturing. But the breakup with Kim has encouraged him to think more like a pessimist. There's nothing tying him to Charlottesville anymore except his buildings, and those can be sold—even the factory. He can live anywhere, build his company anywhere. It's utterly characteristic of Oliver to buy a factory and then daydream about freedom; the way he alternates between brick-and-mortar transactions and visionary speeches is one of his defining contradictions. He says he's beginning to explore manufacturing opportunities in China. "I can make this thing fly in a country like China," he says. "I don't want to go there right now, but I can do with a million dollars in China what a hundred million won't do in America. I can sit in a factory and teach people. I can get people enthusiastic." Oliver says the other employees at Edison2 are willing to consider making the leap.

"If this thing doesn't get traction here, we are all prepared to move to China," he says. "Bobby said he would go. But you know, they are race-car guys, and race-car guys go where the business is."

THROUGH THE end of 2010 and the beginning of 2011, Edison2 is stuck in a holding pattern. It's a time of phone calls and e-mails, meetings and speeches. The men give talks here and there, to luncheons and chambers of commerce. Oliver meets with Steve Rattner, Obama's "car czar," and tells me afterward that Rattner "basically left saying, 'Look, if there's anything I can do for you, I will do it.'" Rattner told Oliver he manages New York mayor Michael Bloomberg's money, seeming to hold out the promise of an investment. However, "like with many of these meetings," Oliver says, "there's no immediate action point."

Brad and Ron go to the White House to brief several members of the President's technology staff. A guy from the body structures department of GM calls Oliver to ask if he can visit the shop. Someone from the venture capital firm Kleiner Perkins sends Oliver an e-mail asking for more information about the technology. It seems like everyone wants to take a peek at Oliver's operation. But that's all they want—a peek.

"We've had a lot of little breaks," Oliver tells me. "It's still a struggle. But it looks to me like we're starting to get some traction. Financially, relatively *dry* traction. But that's what happens. You have to build it."

In the meantime, Oliver is running out of rope. After paying back his investors and suppliers, and giving bonuses to his employees, there's hardly anything left of the $5 million Prize purse.

In mid-November, I get a call from David Brown. He says Oliver is putting twenty-three of his buildings up for sale—twenty-one in Lynchburg and two in Charlottesville, though not the steel foundry he just bought. "He wants to simplify his life," David says, inviting me to attend what he calls "Oliver's event." A few days later, I arrive at the appointed time and place—a modern-American restaurant in downtown Lynchburg—to find Oliver giving a speech to about eighty well-dressed businesspeople. Many of them hold a printout headlined:

> KUTTNER PROPERTY OFFERINGS
>
> ## A CALL TO ACTION

"I am trying to start a movement here," Oliver says, his voice booming through a microphone. "But I can't do it alone."

Oliver says he needs the help of the gathered Lynchburgians. The economy is frozen. The government isn't going to storm in and rescue a place like Lynchburg. "We have to rescue ourselves," he says. The banks have to start lending money. People need to use the loans to start new businesses. "We all need to pick up the rope," he says.

It's a bravura Oliver performance, candid and personal and

self-contradictory: Oliver is encouraging people to invest in the same city where he's trying to unload buildings. Maybe he feels guilty about divesting himself and wants to be sure others pick up the slack.

"If you are interested in owning an occupied building," he says, "I ask you to bid on some of mine. I love my country and I'm not leaving, okay! I just bought a 157-acre white elephant." He means the steel foundry. "But I think this frozen condition is not going to work. It doesn't work for my plan. I am in too much of a hurry."

Oliver sets down the mic and works the room for a time. The Lynchburgians eat canapés. I meet one of Oliver's protégés, Tony West, a young real-estate entrepreneur wearing a white shirt and tie. West says he never even considered real estate as a career until he met Oliver a few years ago. Now he owns fifteen to twenty properties, including this restaurant. "He created me," Tony says of Oliver. "He created twenty or thirty others like me. I guarantee you that he inspired a couple more."

As the evening wears on, the room thins, and I find P.K., the Edison2 crew chief, sitting at the bar, dressed all in black, drinking a vodka tonic with a slice of lime. I ask how he feels about the competition now that it's over. "There are so many fucking parts in a car, I sometimes think we'd be better off building ring-pull cans," P.K. says. He means the soda cans with the metal tabs on top. "How many fucking billions of dollars in patents on that, right? And it's still a formidable engineering exercise. And all you're making is the fucking top of the Coke can." He sips his vodka tonic. "But for the competition, the car's been good enough. Sufficient to satisfy the rules, yeah?"

A few minutes later, Oliver sits down at the bar with a group of his employees that includes Ben Jonas, the formerly homeless train-hopper. Ben is rawboned and has long brown hair. He wears stained corduroy pants and a dark green chamois shirt. Oliver has always admired Ben. He's a man who doesn't need much. He lives simply and burns little fuel. He understands the virtue of lightness. When you are very light, you enter a different realm.

"The car has tremendous philosophical lessons," Oliver says, sipping a Bud. "The car is teaching me a lot."

THREE MONTHS later, in February 2011, I hear from Oliver again. He says he still hasn't signed a deal to manufacture the Very Light Car, but he's moving forward on the next version anyway. He's made some progress, he says, and invites me down to Virginia to see.

The night I arrive, I meet Oliver, Barnaby, Brad, and Ron in Charlottesville. They're having dinner with a documentary team from The Henry Ford museum in Dearborn, Michigan, which has decided to include an exhibit on the Very Light Car. Barnaby has just flown in from California. Since the Prize, he's been working on the design of the new car. But this time, Barnaby isn't the only one with input into the car's shape. Oliver recently hired two California consultants to make the car less spartan—warmer, more human, more approachable. Jason Hill is an auto designer who used to work with Aptera on the 2e; he also helped shape the Porsche Carrera GT and an early prototype of the Smart car. The second consultant, Peter Barnett, is an industrial designer with the defense contractor Northrup Grumman, the same company Barnaby works for. Hill and Barnett will be in Virginia tomorrow. Oliver and Ron plan to show the California guys something they've been working on in the Lynchburg shop: a full-sized, nondrivable mock-up of the new car, made of spray foam and Bondo. The model has been painstakingly crafted over the last four months by Bobby Mouzayck.

At dinner, Oliver tells the Henry Ford people that the new version of the Very Light Car will include additional safety features, like a front bumper at the required height, to bring it more in line with federal standards and customer expectations. But according to Oliver, his engineers have managed to make the car safer and more broadly appealing without sacrificing efficiency. "We actually have the hope that we end up with a *better* car than the X Prize car," Oliver says.

I jump in: Why not sacrifice efficiency to make the car more marketable? Couldn't Edison2 make, say, an 80-mile-per-gallon car that has cup holders and air-conditioning and all the creature comforts the American heart could desire?

"We're competing with some damn good engineers," Barnaby replies. He means the engineers at major automakers. "We're not gonna beat them by being slightly less mediocre."

"There's only one guy who's the best," Oliver says. "And that's what we are striving to be." His goal is still to convince a large automaker to write him a check, like Daimler and Toyota did with Tesla. "And they have to write a check because they trust us more than eighteen hundred of their own engineers."

"That's really the answer to your question," Barnaby tells me. "Why head for a less ambitious goal? It's inelegant."

Oliver tells Barnaby, "You're gonna really like what you'll see tomorrow. It's really beautiful."

"Bobby has worked very hard," Ron says to Barnaby. "Very hard. So please temper your remarks with that in mind."

"I'll be my usual reticent self," Barnaby says, smiling.

"I'm actually burning to know what you think," Oliver says.

THE NEXT morning, the men of Edison2 and the two design consultants from California gather in Oliver's Glass Building in downtown Charlottesville. Edison2 has set up a satellite office there, mostly to pitch investors. One of the X Prize cars is wedged into the lobby, and the walls are lined with glamour shots of the car in motion, sun glinting off its deviant angles.

The museum people set up a camera to interview the team members. Oliver introduces the interviewer to the consultants, Barnett and Hill. Oliver says of Barnett, "One of the things that Peter is really exquisite at is designing cockpits of machines that are life-or-death. *Shit. Has. To Be. In the right place.*" Barnett tells the interviewer he recently worked on the cockpit of a fighter plane, and when he asked bomber pilots, twenty-year-old guys, what they wanted in a cockpit, they almost unanimously said an iPod dock. When it's Hill's turn, he says his job with Edison2 is to "keep the fantasy" out of the car, and to make sure that "when you approach the vehicle, before you even touch it, you have a natural affinity to it. It looks real. It looks handcrafted to a very high level."

Arranged on the long conference table are three scale models— one that depicts the old, X Prize version of the car, and two that depict new variants. The most noticeable differences are the addition of a front bumper and a change in the design of the wheel pods. Now the two wheel pods in the front appear connected to the body and to each other by what Edison2 is calling a "wing." A similar structure in the rear mimics the front wing. Also, the wheel pods are cambered out a little farther, widening the car's stance slightly. It now looks less like a bird and more like a fish—a skate or a ray, or even a hammerhead shark.

It's time to discuss the scale models. Ron opens his laptop and loads some drawings that the consultants have made. Barnaby starts by asking some questions about how the new wing shapes will affect the car's cooling. Within a few minutes, though, the meeting evolves into an extended back-and-forth between Barnaby and Bobby, starting with an argument about side mirrors. Barnaby doesn't want the mirrors where they are on most cars—mounted to the body. Any piece that protrudes, he says, will mar the aero. He'd prefer a video camera connected to a display on the dash. The problem with that is U.S. law, which requires side mirrors.

Oliver says he thinks the car should address the law as the law is written.

Barnaby wonders if they could mount the mirrors on the wheel pods, farther away from the car, to minimize aerodynamic disruption. Bobby counters that the farther away the mirrors are, the bigger they have to be.

Again and again, Bobby proposes an idea to make the car more user-friendly, and Barnaby shoots down the idea on aerodynamic grounds. For instance, Bobby proposes simplifying the wheels by removing the pods. "My Chevy truck, when it rolls off the assembly line, it's ready for Alaska or South Florida," Bobby says. Grandma needs to be able to change a flat tire, and Grandma isn't going to want to mess with a wheel pod. "If you want this car to be cheap and you want to pimp it to the masses, you need to keep it simple."

"Somewhere between that and having the aerodynamics of a Hummer," Barnaby says.

Bobby mentions the "financial burden" of a complex car; the more moving parts, the more expensive it is to make, the more it costs the consumer.

"Well," says Barnaby, standing now, holding a cup of coffee, "that's something I'll discuss with Oliver."

Oliver presses both hands into his temples as Barnaby and Bobby go at it. He stays mostly silent, watching them with a look of intense concentration. Later, when I ask him why he didn't jump in to guide the discussion, he'll say, "Sometimes, if you fight about it for a few hours, you come up with a new answer." In the moment, though, it seems as if no answer emerges—no consensus on which of the new versions are best or which compromises the team is willing to make.

Someone cracks a joke about how they should all be proud that they've figured out the "Alaska version" of the car.

Barnett jokes: "Where do we put the gun rack, then?"
There's a pause.
"Next order of business?" Bobby says.
"Lunch?" Barnaby says.

WE CARAVAN to a bagel place and get sandwiches to eat on the hour-long drive to the shop in Lynchburg. I ride with Ron, who didn't have much to say earlier. As we pass tiny auto-body shops and rural gas stations, the terrain growing hillier, I ask him what he made of the argument between Barnaby and Bobby. "Personally," Ron says, "I don't think this car has to be everything to everybody. I don't see anybody in Alaska buying this car, except as some sort of eco/political statement. That said, making it easier to work, a bit more finished, is a proper objective."

Seventy miles later, we arrive at the factory in Lynchburg. Inside, Ron joins the rest of the group on the main shop floor. They all gather around one of the X Prize prototypes—the original silver car that became a side-by-side coupe. Its fiberglass panels have been recut to conform to the new shape. The car now looks more like an airplane than ever—pointier, more dramatically angled and sloping in the front. The windshield has been removed. Hill climbs into the driver's seat and looks around.

Oliver says, "I look at you right now—you look like you're in a jet plane. You look really cool. I think it's just a matter of getting used to it. And I think you can turn it around into, actually, 'I really want one of those cars that looks like a jet plane.'"

There's something else about the car that's different. Under the hood is an empty oilcan. The can has been painted with the words "ELECTRIC MOTOR." This means that in addition to working on the next model of the gasoline-burning Very Light Car, Edison2 is also working on an electric version. Oliver has hired the team leader of Li-Ion Motors, one of the other X Prize winners, to work out the details of the motor and the battery pack. (By all accounts, he's a capable guy, not guilty of Li-Ion's sins.)

At first, I'm confused by this development. Oliver is the guy who won an electric car–dominated contest with an internal-combustion-engine car. He spent years telling everybody that electric cars weren't as good as they seemed. But the more I think about it, the more an electric Very Light Car makes sense, because it will bolster Oliver's main argument: It's not the energy source that matters, it's the platform. Lightness. Aerodynamic efficiency. Oliver must have seen that an eVLC would be an irresistible response to his critics. *What, you don't like our car because it burns gas? Well, here's an electric version. We made it in a couple months. No biggie.* The eVLC will eventually post a startlingly high MPGe of 245 in EPA-accredited tests, with 114 miles of range, its low weight and aerodynamic efficiency allowing it to best the 2011 Leaf (99 MPGe) and the Volt (93 MPGe) even while using a smaller battery pack and a less powerful motor.

The group walks back to the body shop, where Bobby has created his full-scale foam model of the Very Light Car 4.0. It's actually two cars in one, split down the middle; on one side is the preferred shape of Barnaby, Barnett, and Hill, and on the other is the shape that the men in Lynchburg liked a bit more. There's a slight difference in the roofline and the shape of the wings. "It's an athletic-looking car," Bobby says.

"It's amazing how useful it is to have this model here," Barnaby says.

"Oh, it's nuts, man," Bobby says. "Touch it, feel it, sniff it."

For the next two hours, the men examine the foam model, circling it repeatedly, running their hands over it, murmuring thoughts about the wings, the wheel pods, and how to test the aerodynamic qualities of the new shape. P.K. stands at a remove, scowling. He leans into my ear: "How long are they going to talk about this? And there's still nothing definitive. I mean, how long do they think we have to make something people like? Why don't they get one of those marketing focus groups? Even while we're designing something, nobody really knows if it's ready for the great unwashed."

A little before 5 P.M., I follow P.K. and Bobby outside for a smoke break.

"They're just jerkin' off in there," P.K. says.

Bobby laughs. He says he respects the California consultants, but they're still a little too geeky for his taste. "What we really need," Bobby says, "is a guy from Hyundai coming in and saying, 'No no, no profit there.'"

Back inside the shop, the engineers agree on a course of action and tell Oliver their plan. Edison2 should build four slightly different small-scale models and test them in the academic wind tunnel at Virginia Tech. Then they'll pick the best-performing model, make a full-scale version of that, and return to the wind tunnel.

Oliver agrees. But since he has all of these smart people here, and he's not sure when they'll all be back together, should they all spend some time talking about the shape of the windows?

It's dark outside when the consultants finally pile into a van with Oliver, Ron, Barnaby, and Brad. The van leaves for Charlottesville. Bobby and P.K., who live in the factory, stay behind.

They walk around to the back and knock on a door. A man inside tells them to come in. It's Harry, their pal, a carpenter who keeps a shop in the building, complete with a beer-stocked fridge. Bobby and P.K. refer to Harry's shop as "church." Harry is their pastor. They sit down, drink a few beers—which they call "saying prayers"—and think about Oliver.

Oliver won the X Prize. He mastered an artificial universe, a bubble where the normal rules of commerce don't matter. Now the bubble has popped, and Oliver has a payroll to make and a company to keep alive. But he still seems preoccupied with making a better car.

"We just want to see Oliver done right," Bobby says. "Because he really has put his heart into it. But I've been on two race teams that burned through all the money. And when the money was gone, there was nothing left."

OLIVER IS one of the most open people I've ever met. I've always felt that I can ask him anything. But I hesitate before asking him about the mechanics' case that what the car really needs is a focus group. It feels almost cruel. The Very Light Car is a car that is also an argument about why other cars fall short. They fall short because they're designed around shifting human desires instead of absolute physical virtues and inescapable social needs. Oliver vowed to design a car the other way around. This was the scratch he started from, five years ago, egged on by a $10 million bounty and a space kid he'd never met. And then everything grew from that. The passion for lightness. The toil in Lynch Vegas. The faith in the judgment of his guru engineers. To ask Oliver about a focus group at this point would be to question everything, including the relevance of his victory in a competition designed to make focus groups look like part of a world gone by.

But I ask him anyway: Why hasn't he done a focus group?

"You know," Oliver says, "the Volkswagen Beetle was not designed to be appealing to people. It was designed to meet a task. I think it's Steve Jobs who said, 'Ten years ago, nobody knew they needed an iPhone.'" Oliver pronounces *Jobs* as "Yobs."

His task, he says, is very simple: make a car that uses less energy, is safe, is fun, and looks good. "If we succeed in doing this elegantly, I believe the car will sell itself."

"But don't you think the world is full of elegant ideas that don't catch on?"

He swings his head to the side, as if looking for the punch line to the joke.

"No, the world is *not* full of elegant ideas," he says. "It's actually scarily empty. You name *one* piece of hardware that has the power of this."

Listening to him, I get an image of a man in a hermitage, bent in prayer. An acolyte of a nameless god. Oliver is waiting for the deal he knows is out there. There are large and mysterious forces afoot in skyscrapers and boardrooms, and they are watching. They exist on a different plane, a plane he can't reach, but they are there, and they could descend at any moment and change his life forever. He is being watched, and judged, and he does not want to be found wanting.

He made this clear to me, in a more personal way, two nights earlier, when I first arrived in Charlottesville and stayed with him. Ever since he and Kim separated, he'd been living in the guest-house behind his actual house, the place where the family's au pair had been living. Now the au pair had moved into the main house, with Kim and the kids, and Oliver lived here.

The place was mostly bare, although Oliver had put up some photos of himself in college, rowing for Boston University. He'd begun to line his bedroom with car memorabilia and his extensive collection of car books. He had also created a special bookcase containing more than thirty books by his biological father, the archaeologist and explorer.

Oliver set me up in a spare bedroom with a space heater. There was a child's slot-car track on the floor. He sat on the edge of the bed with his hands in his lap and said he was officially divorced now. He had just signed the papers. Kim was already engaged to her new boyfriend. "The truth is, I do miss living over there with the family," Oliver said. "I like that." He looked out the window and could see lights on in the main house, thirty feet away.

"You know what the real underlying thing is? You work really hard and there's no money made. It's a lot more fun when you work really hard and sometimes there's a check." He sighed deeply.

"The car work? I really *like* it. I like it. And I'm just not that interested in my real estate anymore."

He considered this for a second.

"You know, a lot of people in the world don't like their jobs. . . . I shouldn't complain. Besides, we have a slot-car track, and that's what matters." He pointed to the toy on the floor. "The fact that we have a slot-car track puts us in the top 1 percent of the population. This is a good problem to have, on the floor."

Oliver walked out into the hallway and came back brushing his teeth. He said that one of his goals now was to make enough money to provide for Kim and his kids in perpetuity. He said he had recently taken out a $2 million life insurance policy, because "something that happens to guys like me is we have heart attacks at age forty-eight and drop dead."

He returned the toothbrush to the bathroom and sat back on the bed.

"If I really hit the jackpot," he said, "that girl won't have to worry about money, ever. And I might marry another woman and have another four kids."

Oliver's voice was calm and remote, as though it was coming from some place far away and he was trying to convince himself. He had seen so many couples fail, he said. He had seen so many car companies fail. He wasn't going to try to be a car company. He wasn't going to drag the kids through the mud. He was going to be fine. Whatever happened, he was not going to be lonely.

25

The Old Frontage Road

After the winner's ceremony in D.C., Kevin Smith returns to Illinois. He sits in his cubicle at the EPA, surrounded by *Star Wars* figurines and newspaper clippings about Illuminati.

No one from the auto industry has offered him a job. No one with the power to make his car real has shown the slightest interest. Despite having proven itself on the track, the car is so different as to be invisible. *Since they can't conceive of it,* Kevin thinks, *they can't do it. Why would an industry that's made trillions of dollars doing it their way do things any differently?* He consoles himself with the thought that if some electric-car enthusiast falls in love with the car and decides he must have one, Kevin and his team could build a single copy for $250,000. To get the price down to $42,000—about the cost of a Chevy Volt—Kevin figures he'd need $40 million in capital.

During the Prize, sixteen-hour days in the shop felt effortless, but now, at the Illinois EPA, every minute working on a pollution permit is excruciating. Kevin's mind keeps wandering back to the contest.

He was so busy for so long that there wasn't time for broader doubts to creep in. In his private journal, in a long entry titled "The Project," Kevin grapples with what it all meant. The contest felt a little "like growing up all over again," he writes: the "schoolyard bullies," the joy of discovering a power within yourself, and the "hurt, confused feelings" sparked by failure. Much of the entry is self-flagellating. Kevin chides himself for "my nearly limitless capacity of ignorance" and writes that it's possible the entire Illuminati effort was built on the quicksand of "an unfounded, unrealized, very momentary belief in myself." All the same, he tries to find something positive in the loss. "The only thing worse than losing would be winning," Kevin writes, because losing "gives you desire."

In the journal, he also does something he hasn't really done in years: He takes stock of his relationship with his wife.

Kevin knows that, during the Prize, he was partly absent as a husband. He's not sure how Jen kept the household together while he was in the shop day after day, trying to prove that he had something important to offer the world. "She must have seen a man digging a trench by hand," Kevin writes, "in order to stop the oncoming flood."

One night he and Jen go to a double feature at a drive-in theater off Route 66. Instead of taking his Pontiac Vibe, Kevin decides to take the X Prize car. By now, Seven is fully street-legal and insured, as legitimate in the eyes of the law as any car on the road; Kevin got a title and had it inspected, and there's a registration sticker on the license plate, which reads IMW 77. Nate, Thomas, and their families follow in separate vehicles. The car ends up drawing a crowd before and after the movies. Men and women circle it, gawking, and Kevin lets a couple of kids sit in the front seats. As the members of Team Illuminati answer questions, a heavyset man comes up and asks what kind of car it is. Kevin says it's an electric car.

Seven at the drive-in.

"Yeah," the man asks, "but what's it run on?"

"It's electric," Kevin says, smiling.

"Yeah, but what's it run on?"

"It runs on batteries. Ninety-six batteries down the length of the car. You can see the battery box here."

"Yeah, but you say it gets 182 miles per *gallon*," the man says, gesturing toward the car, where Jen has taped a custom sticker to the passenger-side window. It looks like the ones you'd see at any car lot, and says the car's highway fuel economy is 182 MPGe, based on its performance at the X Prize.

"Miles per gallon *equivalent*," Kevin says. "In electricity."

"So what's it run on?"

Kevin pauses. "Ethanol," he says with a straight face.

"Huh," the man says. "Ain't that something."

Around this time, Josh Spradlin tells Kevin he should pick up a fantasy novel called *The Magicians*, by Lev Grossman. The book is about a boy who grows up reading novels about a fantasy land called Fillory and then discovers that Fillory is real. When he visits Fillory, though, it's not at all how he imagined. Kevin reads *The Magicians* and copies a passage in his journal:

> Look at your life and see how perfect it is. Stop looking for the next secret door that is going to lead you to your real life. Stop waiting. This is it. There's nothing else.

ONE EVENING in December 2010, Kevin is invited to give a speech to a young engineers' club in Springfield. He brings a PowerPoint presentation about the Illuminati car, but when he tries to use the projector, it breaks. So he just stands there and talks.

He talks for two hours straight.

At the end, Kevin says that if there are any kids in the audience who are interested in electric cars, they're welcome to come out to the shop and take a look.

A woman approaches him, followed by a shy sixteen-year-old named Matt with braces and a mop of brown hair.

"As long as he doesn't lose any fingers or get electrocuted, you can have him," the woman says.

The next weekend, Matt drives out to Kevin's property in a blue station wagon. Kevin gives Matt a tour of the shop and gestures to a broken transmission sitting on a table.

"Have you ever torn apart a transmission?" Kevin asks.

"No," Matt says.

"Neither have I," Kevin says. He hands Matt a wrench. "Here you go. Holler if you have any problems."

Four hours later, the transmission is all taken apart and labeled, nice and neat, and there's only one loose ball and spring on the floor. Kevin picks up the ball and spring and says, with a stern face, "What's this? It looks like crap." Then he cracks into a grin and tells Matt he's doing great.

To Kevin's surprise, Matt returns the next Saturday, and then the next. He seems to think the car is something amazing.

Is it?

At the Prize, Seven racked up a respectable combined fuel economy of 119.8 MPGe. But that was with a junk transmission, a Slip 'N Slide clutch, and a buggy electrical system. What could it do if the problems were fixed? What was Seven's real potential?

Kevin knows a way to find out. The X Prize has made a deal with the U.S. government. The government uses a particular test to peg the fuel economy of electric cars like the Volt and the Leaf. Any Prize team can now bring its car to the Chrysler Proving Grounds in Michigan and have it measured by the same official yardstick.

The deadline is March 2011—three months away. The Illuminati car is spread out across the shop, in pieces.

Kevin reaches out to Nate, Josh, George, Thomas, and Nick. Soon they're all in the shop again, every weekend, furiously building. Everyone, that is, except Jen.

RIGHT AFTER the new year, I e-mail her with some questions about the car. "I couldn't tell you," she writes back. "In a vain attempt to recover my former life, I quit the team, only to discover that my former life only exists as a concept. Please direct all IMW questions to Kevin. There's a lot I'm sure he'd love to talk about."

It's a purely strategic move. She can't deal with another intense burst of building. She can't do it again like she did it before. She needs more control over her time, her life. She decides to give up her role as an official member of Illuminati.

She doesn't cut the cord entirely, though. Jen stays attached to the team as an occasional volunteer. She realizes that if she wants to be married to Kevin, she has to maintain some kind of relationship with the car.

A few times during the contest, she did think about leaving him. But she never got far. She kept bumping up against the high wall of their shared history: the summers they spent together as kids in Park Forest, the college years they endured apart, their life and debts in Divernon. And she didn't want to interrupt the project and kill a dream.

One day in early 2011, Jen sits down with her husband and tells him how she's been feeling. It's their first extended personal discussion in almost two years.

"Reintroduction," she reports on her Twitter feed. "Communication. Combined future plans. Hope."

BY THE end of March, Kevin and the team have added three new batteries to the car, stripped redundant steel plates from the doors, recarpeted the entire interior, modified the transmission to remove first gear, second gear, fifth gear, and reverse (the motor itself has a reverse mode, so the reverse gear is redundant), and completely rewired the battery pack and dash. They've also removed part of the clutch assembly and bolted other parts together. This means they can no longer shift gears while driving—they have to start in either third or fourth gear and stay there—but it also means the clutch will never slip again. The car now weighs 2,900 pounds, nearly 250 pounds less than its Prize weight. Kevin and a few of the guys haul the new and improved Seven back to Michigan, to the Chrysler Proving Grounds.

At Chrysler, Kevin hands the keys to the company's technicians, who perform a Coast Down test. Then the Chrysler employees strap Seven to a dyno (a treadmill) and perform the EPA's official "test cycle," driving the car to simulate both city and highway usage. Kevin and the guys wait for the results in the lobby of their hotel, watching Fox News and *All My Children*.

Eventually the Chrysler people hand Kevin a stack of papers. There are a whole bunch of numbers on them that describe the car's performance. Later, he plugs the numbers into the EPA's formula to get the official EPA fuel economy of Seven.

When Kevin gets home, he checks around to see if he can find a street-legal sedan with a higher number. He can't. Then he tries to find *any* street-legal vehicle with a higher number. He can't.

The 2011 Chevy Volt, according to the EPA, gets 93 MPGe. The Nissan Leaf gets 99. The Foundation measured the two-seat Li-Ion Wave II at 187, and the two-seat Aptera 2e at 179.3. And the four-seat Illuminati Motor Works Seven? A handmade car, built of fiberglass and melted steel, in the middle of a cornfield, by a team of amateurs? Seven gets 207.5 MPGe.

Kevin e-mails the number to Jen, who gets to work designing a new window sticker. Prize officials have given each team a flyer that compares their car to a Prius, and Jen includes the information from this flyer on the back of the sticker. Seven is 35 inches longer, 10.7 inches lower, and 142 pounds lighter than a Prius.

207 MPGE, it says in a bold font at the top of the sticker.

Annual Electric Cost: $264.

Development Status: Prototype.

I VISIT Kevin and Jen one afternoon in October 2011, a little more than a year after the end of the Prize. The corn in Central Illinois has just been harvested, and a mist of dust and particulate plant matter hangs above the field across from their house, filtering and intensifying the sunlight. When I pull into the driveway, I see Kevin and Jen standing outside, next to Jen's AMC Gremlin. Jen has just put a new seal on the passenger-side door and now the door won't shut. Kevin is trying to force it shut. He has tied one end of a rope to the door and the other end to a ratchet that will tighten the rope. He holds the ratchet against the side of the car.

"Rag," he says, calling for a rag.

"Rag," Jen says, handing him one.

"Just hold it right here," Kevin says. He's trying not to scratch the paint job.

Jen says that's not necessary.

"It's gonna mess up your paint," Kevin argues. "And then it's gonna rust."

"That's what Rust-Oleum is for."

Kevin cranks the ratchet and tightens the rope. He stands back and appraises the door.

"We're gonna have to shim it," he says. "The car didn't come with shims."

"Didn't come with rust, either," Jen says.

"Best Detroit has to offer," says Nate Knappenburger.

Today Nate is his usual placid self—big Nate, formerly of the U.S. Air Force and ITT Tech; Nate, who wired up a 207 MPGe car and acted like it was no big deal. He tells me he sent out some résumés after the Prize, but nothing came of it. He did get contacted by Tesla Motors, and they flew him out to California for an interview, but Tesla eventually went with another candidate. "They can't afford not to hire the absolute best in the industry," Nate says. "All I have as a credential is the X Prize." He shrugs. "It was an honor to even be considered."

Kevin gets the door working and we all go into the shop. There's Seven, gullwings up, hood up, paint shining. The array under the hood looks newly tidy, almost minimalist—no wires spilling everywhere, just a couple of boxes and hoses. Kevin reaches for a cardboard box beneath a nearby shelf and hefts it up to neck level. It's full of steel plates. "We took all of this steel out of the car," he says. "We didn't need it." There are also some nice user-friendly touches: Kevin has cut little circular vents in the driver's-side and passenger-side windows to let cool air into the cabin, and Nick has added cup holders. They aren't done yet. Future improvements are listed on the whiteboard, including:

> New Trans
> New hood
> New Doors
> Drip edges
> Reshape Front & Rear Fenders
> New Dash
> Graviton conduit
> Anti tachion displacement emitter AKA flux capacitor
> Remove "Flood Hole"
> New radiator . . .

Graviton conduit. Kevin is thinking about gravity again. A good sign for his mental health. It means he's back to his old self.

"Have you seen the aluminum?" he says, and takes me into the shop's storage room. On the floor is a waist-high stack of long aluminum honeycomb material. "We can make seven cars out of that," he says, beaming. "Full-size."

The next version of Seven won't be made out of fiberglass and shaped by hand; instead, it will be made of this high-end composite stuff.

Kevin glances at Seven, then at me.

"So, wanna go for a little ride?"

We climb in: Kevin and me in the front, Nick and George in the back. We shut our gullwing doors. Kevin pulls out of the driveway. Gravel spins beneath the wheels. The odometer reads 5,277.

He takes a right on Montgomery Road. Corn on the left, soybeans on the right. "We seem to have less problems with it now," he says as he accelerates up to 45 in a couple of seconds. There's a crunch. "That was a beer can."

He slows and pulls onto the frontage road, the old Route 66, which parallels I-55. He points the car south. Ahead of us are eight miles of arrow-straight asphalt—no other cars in sight.

Seven lurches forward. Ten seconds later the speedometer reads 66. Acceleration is what doomed them in the Prize, but now the car accelerates quickly. The ride is jerky; with each bump in the road, the front of the car seems to lift off the ground slightly. "You feel that little bounce?" Kevin says. "That's because we're an ellipse. The car is trying to take off." Seven makes a noise like a Tuvan throat singer, a low rumbling mixed with a higher, harmonic pitch.

It's now at 77 miles per hour.

"So this is our test track," Kevin says, with a touch of bravado. "We'll see how fast it can go."

The car shakes . . . 90 miles per hour.

In the backseat, Nick and George look out to the left side of the car, at the drivers heading south on I-55, parallel to Seven . . . 100, 110 miles per hour. The sun going down. The horizon aflame . . . 125. A new shape cutting across the band of orange at the joint between corn and sky. Kevin shoots me a glance. He's grinning maniacally. The drivers on I-55 look to their right and see four men arrowing across the newly harvested cornfields at 128 miles per hour. Past silos, past tractors. The I-55 drivers wave at the men and the car, give the thumbs-up sign. They watch the car narrow into a pale silver dot. Then there is only the harvest debris, the threshers, the bare brown fields.

epilogue

Airborne

pull into the Philadelphia Navy Yard one morning in May 2012. The building I've been told to look for, an old Victorian painted bone white, is next to the water, with a view of ships in the harbor. I know I'm in the right place when I see a box that reads MITER SAW WITH LASER next to a wicker chair on the front porch.

I walk through the door of a sunroom. A poster on the wall says:

WE ARE THE SUSTAINABILITY WORKSHOP

We will be known as . . .

Intelligent	Motivated
Resilient	Observant
Responsible	Respectful
Innovative	Humble

"Hey, how's it goin'?" Simon Hauger says, extending his hand. "I thought one of the kids could give you a tour."

This place, the Sustainability Workshop, is the closest Simon has ever come to running his own school. Late last year, he convinced the Philadelphia School District to let him design a project-based curriculum for a handful of seniors. Twenty-eight kids from three high schools signed up. Instead of attending their usual classes, they spend their days here at the Navy Yard.

Anthony Tran, an eighteen-year-old in glasses, shorts, and high-tops, offers to be my tour guide. He's soon joined by another volunteer, Rasheed Bonds, seventeen. As Anthony and Rasheed walk me through a series of high-ceilinged rooms flooded with natural light, they describe the way the school operates. The students use a software package called Project Foundry to define what they're working on and how they're going to finish it. Each week they have a "deliverable," something concrete that they and their teachers agree is due on Friday. They track their own progress. There are no classes to speak of and no classrooms. Whenever a group of students begins a project, the group chooses a room. The students move freely, carrying Google Chromebooks.

Upstairs, they proudly show off the school's Makerbot, a machine that prints 3-D objects, and in the backyard, they point out the school's vegetable garden, where the students are growing lettuce, peas, and Swiss chard. Near the garden, next to a shed, is the body of a very familiar-looking car. From the top it looks like a diamond. From the side, a bird skull. It's unmistakably a Very Light Car. Oliver Kuttner, it turns out, has donated one to the school so that the kids can swap in an electric motor. (The West Philly Hybrid X Team lives on at the Sustainability Workshop, but it's much smaller now. Several key members have moved on since the Prize. Jacques Wells, for instance, completed his first year of college at Penn State–Berks, then had to get a job because he lost his financial aid. He now works at a baking facility in Reading, Pennsylvania.)

After the tour, Simon and I drive to a nearby coffee shop. "This is my best year professionally ever," he says. "It's worked better than we imagined. It's kind of like building a car. You know there are going to be major problems. And we haven't had *any* major problems. And maybe it's because we've been prototyping and testing all this time. All the big pieces are working really well. There's tinkering that needs to be done. But the outcomes are incredible. Every one of those kids will tell you that they've done more work than they've ever done in the last three years, cumulative, *and we're not directing it."*

I shift the conversation to cars. I'm curious to know how Simon feels about the X Prize, looking back. Of all the teams that were serious contenders, his team probably got the rawest deal. West Philly listened closely to the X Prize's early—and, as it turned out, faulty—signals. They were told the winner would be determined by a cross-country race, so they traded some fuel economy for raw speed; they were told the cars had to be "real," so they made sure theirs were affordable. They paid the price, missing the finals by just three-and-a-half MPGe.

Yes, the rules changed in midstream, Simon tells me. If he had it to do over again, he would have designed the cars differently. "But so would everybody else in the competition." In the end, it doesn't matter that West Philly fell short of the finals. They still beat ninety other teams, including Cornell.

"They took us seriously," Simon says. "We're tinkerers. We're not legit!" He laughs. "And they gave us a venue to a world stage."

The experience hardly scared West Philly away from competition. In 2011, the team entered a race for alternative-energy vehicles called the Green Grand Prix, in Watkins Glen, New York. More than two dozen teams brought cars to Watkins Glen. West Philly drove its GT sports car for more than 100 miles, and at the end, when all the fuel was measured and weighed, the car posted an MPGe of 160, more than good enough for first place, beating a Chevrolet Volt, which got 129 MPGe. West Philly's mileage was so high because the calculations weren't as rigorous as they had been at the X Prize. Still. In 2012, the team returned to the Green Grand Prix and successfully defended their title.

In 2013, a year after my tour of the Sustainability Workshop, I get word that Simon has convinced the city of Philadelphia to let him scale up the Sustainability Workshop from a pilot program to a full-fledged district institution. The plan is to move from the Navy Yard into a building in West Philly. Simon will finally, at long last, get to run his own school.

APTERA MOTORS didn't prove so resilient. It shut down in December 2011. It ran out of money waiting for a government loan that never came.

In interviews with journalists, CEO Paul Wilbur partly blamed himself for the company's failure. He said he had succumbed to "bright shiny object disease" in chasing the loan so intently for so long. But he also complained that broader political and economic trends had made Aptera's job impossible, particularly a controversy surrounding a formerly obscure solar-energy company called Solyndra. After borrowing $527 million from the Department of Energy (DOE), Solyndra had gone bankrupt, and when it went down the drain, it seemed to drag the DOE's loan program down with it; right-wing radio hosts yammered for days about "Obama's green scam," and DOE secretary Chu was forced to answer hostile questions from Republican congressmen.

The Solyndra affair probably made the DOE more hesitant to promise funds to a small start-up with an odd-looking electric car. But another thing that may have hurt Aptera was the early sales figures of other electric cars. In 2011, Americans bought about 17,500 plug-in electric cars, and the number tripled the following year, to 53,000. It seems likely to double again in 2013. Both the

Chevy Volt and the Nissan Leaf sold more in their first year on the market than the Prius did in its first year. Demand for electric cars was clearly growing. But the numbers were still beneath what the car companies had publicly predicted. A broad market hadn't materialized. Not yet.

On Aptera's final day, a few of the engineers asked if they could drive one of the 2e prototypes for one last time. They climbed in and did laps around the office. When they were done, the nineteen remaining employees gathered around the car for a group picture. Some of the prototypes were saved; the rest were given to liquidators.

NOT FAR from Aptera headquarters, the X Prize Foundation was leaving the auto prize in the past. New prizes, new jackpots: $10 million to sequence genomes faster than ever before; $10 million to build a working "tricorder" to diagnose medical conditions, like in *Star Trek*; $30 million to land a robot rover on the moon. Its influence spread. One senior employee left to join the Obama administration, where she helped build a sort of X Prize program within the federal government—a website, Challenge.gov, where federal agencies can post cash bounties and lay out problems for citizens to solve.

And Peter Diamandis? The man who'd believed that the engineering and fiscal woes of the major automakers had opened up a window in the market—one he could help small entrepreneurs climb through? Well, he was moving on. In 2012, he helped launch a new company, Planetary Resources, to mine asteroids for platinum and rocket fuel. He also cowrote and published a bestselling book, *Abundance,* about how emerging technologies will solve the world's hardest problems. It came wrapped in aluminum foil, a formerly scarce metal made abundant by clever mining techniques.

In the book, he spent pages describing the space prize, but the auto prize merited only a brief paragraph. He didn't mention the winners by name.

It was obvious by now that the auto prize hadn't worked out as the Foundation had hoped. "I don't think it was a bad thing," says John Voelcker of Green Car Reports. "I think it was just hard to explain and irrelevant to the industry." No flamboyant billionaire had built a company around the winning vehicle, as in the case of the space prize; no great paradigms had shifted.

Part of the problem was the way the contest was designed. By ignoring public input, the organizers had guaranteed that the winning cars would be the ones that had been optimized to win, not necessarily to appeal to the driving masses. You could hardly blame the winning teams for winning; all they'd done was conform to the rules, brilliantly. But now, back in the real world, they were struggling. In the summer of 2011, the auto blog Jalopnik checked in with Edison2, X-Tracer, and Li-Ion and concluded that none of them had the money, the capacity, or the relationships to produce their cars. "At this point," Jalopnik wrote, "the Automotive X Prize shows little to no chance of revolutionizing the car world through its science-fair-meets-Great-Race circus."

Yet the X Prize teams were hardly alone in failing to transform the industry. By 2012, virtually every prominent automotive start-up in America was either dead or heading in that direction, even the ones backed by millions in low-interest government loans. It came back to the heinous difficulty of the auto business: the high start-up costs, the regulatory complexity, the entrenched position of the major automakers, the inherent conservativeness of consumers. Bright Automotive, an Indiana start-up that wanted to make plug-in electric utility trucks for government fleets, filed for bankruptcy in 2012, as did Massachusetts battery manufacturer A123. They were followed into the boneyard by Fisker Automotive, whose astonishingly heavy and poorly performing plug-in hybrid, the 5,300-pound Karma, had been panned by every reviewer.

Meanwhile, the major automakers were surging. GM, Chrysler, and Ford were selling cars and trucks and making money again. By late 2012, their factories were operating above 100 percent capacity, up from less than 50 percent three years earlier. In early 2013, Ford announced its most profitable quarter ever. It was as though the Great Recession had never happened. The window in the market had slammed shut.

Of all the hopefuls, only one had squeaked through: Tesla Motors.

In late 2012, Tesla released the $70,000 Model S, the premium electric sedan it had dropped out of the X Prize to develop. *Automobile* and *Motor Trend* instantly named it their car of the year for 2013, and *Consumer Reports* gave it a rare 99 out of 100 rating, calling it the best car they'd ever tested. I got to see the Model S myself when Tesla opened its first dealership in the Philly area, in an upscale mall, next to an Apple Store. It was Tesla's twenty-fourth store in the U.S. and thirty-fifth in the world. The store's metal gate was closed when I got there—they weren't quite ready to receive the public—but a salesperson, a petite blonde in a tan jacket, let me in.

Two Model S's: one red, one white. "Real" cars. Recognizably sedans. But beautiful. None of the weirdly packed stoutness of the Prius, only sleek lines.

She pointed to a bare chassis on wheels in the middle of the store. The Model S chassis, she said, was made of aluminum and designed from the ground up, unlike the Tesla Roadster's chassis. She showed me where the battery pack went, under the floor, and pointed out the electric motor, which she said was "the size of a watermelon," one of the friendliest fruits. The store seemed to encourage hanging out. Touch screens offered facts about electric-car technology, and shelves and walls were lined with Tesla-branded T-shirts, hats, water bottles, lunch boxes, and even baby onesies ("0 emissions baby! . . . almost").

The salesperson walked from the chassis to the white Model S. The body seemed completely flat, with no visible door handles, just outlines where the handles would normally be. There was a mechanical whirr and the handles poked out. We climbed into the car and the handles slid back into the body. (The retractable handles improve aerodynamic sleekness. The Model S's coefficient of drag is .24, not nearly as good as the Very Light Car's .16 but better than almost everything on the road, including the Prius.) I sat in the driver's seat, facing the biggest touch screen I'd ever seen in a car: temperature controls, Internet radio, the works. On the digital dash, a graph showed the car's energy usage. "You'll be able to calibrate your range in this car even better than you could in a gasoline car," the salesperson said.

I asked her how long it would be before a person like me could afford a Tesla. She said they were working on it; first would come the Model X, an electric SUV, in 2014, and then, after that, an electric car for the masses. As we spoke, people walking by the store pressed their faces against the closed gate.

All along, I'd been less interested in Tesla than in the X Prize teams. My sympathies were with the people designing cars I could potentially afford, and it had always seemed like Tesla was making toys for the wealthiest 1 percent. But the company's strategy of starting on the high end and working its way down was turning out to be a smart one. Over its ten-year existence, Tesla had innovated bit by bit, getting cars to drivers in increments, building trust and experience, and growing bolder as it went, but never presenting consumers or investors with anything too radical, too big a departure from what they were used to. And now the company was profitable. In the first quarter of 2013, Tesla shipped 4,900 Model S's. On tech sites and investment boards, journalists and hedge-fund managers began wondering aloud if Tesla was the next Apple Computer, the indispensable company that would make everyone ridiculously rich. Its stock rose from $33 a share at the beginning of the year to more than $100 by July, the month when CEO Elon Musk raised $830 million from the markets and paid back Tesla's government loans ahead of schedule.

In the public eye, Musk was fast becoming his generation's iconic inventor. He was the wizard who'd told the market to fuck off and followed his own muse. Oliver Kuttner had done that, too. But Musk was selling products. He was an Oliver who was making it.

"THERE IS a human condition," Oliver says. "People will preside over their own demise, rather than make a change."

It's January 2013, and we're in Charlottesville, walking from Oliver's favorite coffee shop to his pickup truck. I haven't seen him for about a year. He's wearing a gray houndstooth blazer over a dark green T-shirt. He looks healthy. No weight gain, no extra strands of white in his hair. He tells me a story as we walk.

There's this building he owns—an old warehouse. He rents out an apartment unit in the building. The unit has no window. Oliver wants to cut a window, to make the unit more desirable. But his financial partner, who owns part of the building, knows someone who's willing to rent the unit right now, without a window.

The partner wants to go ahead and rent the unit. Oliver wants to wait and cut the window. "You have to bite the fucking bullet," he says. "I see this movie everywhere. It is not about the partner. It is a human condition. As a country, we're not willing to invest in the future."

Oliver's driving now, one hand on the wheel, heading to the Edison2 shop in Lynchburg to show me the latest version of the Very Light Car 4.0. He swivels his head, looks me straight in the eye, and starts catching me up on recent events, narrating long, giddy paragraphs without paying much attention to the road. The truck drifts onto the shoulder, perilously close to the guardrail. Oliver jerks the steering wheel back just in time.

I gather that life has recently gotten easier for Oliver. To start with, his real-estate business has picked up. He was able to sell about 25 percent of the properties he offered at his "Call to Action." Across all his buildings, he owns 120 to 130 rental units—it's typical that he doesn't know the exact number—and right now only 4 units are empty. Thanks to the broader economic recovery in the U.S., his renters have more money in their pockets, and the banks are trying to lend him money again. "I am feeling the banks come around. I lasted five years through this real-estate shit, and I make more money now, not less."

His debt, he says, is a third of what it was before the recession. "I'm fifty-one years old. I don't want to owe a ton of money when I'm sixty. To get there, you gotta push. Ideally, I owe nothing when I'm fifty-five. I think I can do it."

Oliver survived the worst recession in eighty years. But maybe more important, he survived his divorce. It's now been two and a half years since Kim moved to Prague to be with her new husband. Oliver isn't raw anymore. He has settled into a new routine. He flies regularly to Prague to see his kids. He sits down at the dinner table with the kids and Kim and her husband—"I like him; he's a really nice guy"—and everyone gets along just fine. While it's draining for him to have to travel so far, he says he has made his peace with the arrangement. Kim has her guy, and Oliver has—someone. He won't tell me any details, only that he's not alone anymore, romantically, and he's happy.

Edison2 is a different story. Edison2 is not quite alone. A handful of large institutions are giving Oliver support, including Siemens, the global engineering firm, which provides Edison2's CAD software, and Ohio State University, which has a large Center for Automotive Research and is helping Oliver make parts in casts instead of carving them from billets. Later this year, Oliver will be invited to give the keynote speech at a major automotive conference in Michigan, focusing on the use of lightweight parts.

"The industry clearly sees we have something," Oliver says. But the corporate marriage of his dreams remains elusive. The China deal, the one he was so excited about the previous year, never went anywhere; Oliver says the Chinese were willing to pay him, but he wasn't happy with the "terms and control," so all he has to show for his time is a pen with an ornate dragon design that was given to him as a gift by a man from Guangxi Province. Nor have any large automakers offered to buy Edison2 or license its technology, despite Oliver's tireless evangelism.

What's frustrating about this is that the automakers, in their public statements, have started to sound a bit like Oliver Kuttner. After spending decades pumping cars and trucks full of steroids, they've suddenly become avid about lightness. The U.S. government has essentially forced them to care about weight. According to the new regulations, automakers need to double the fuel efficiency of their fleets by 2025, thereby halving greenhouse-gas emissions. To clear the regulatory bar, they'll have to make efficiency gains across all segments, from compacts to trucks to SUVs. Cutting weight is the most obvious way to do it. This is why Ford, a company long associated with brawn, has lately started to brag about the *smallness* of its engines. V8s are becoming V6s; 6-cylinder engines are becoming 4-cylinders. In 2012, the *Wall Street Journal* reported that Ford was making a major investment in aluminum pickup trucks; starting in 2014, F150s would roll off the line with aluminum body panels instead of steel, shaving 700 pounds from each truck. (Ford later backed away from the article, saying it had no definitive plans regarding aluminum F150s.) Other companies have announced new models containing lightweight materials. BMW is set to roll out the first mass-produced car with a carbon-fiber body, a 2,800-pound electric vehicle called the i3; GM is experimenting with panels made of magnesium. And Volkswagen will soon start offering test drives of the 1,750-pound XL1, a more evolved version of its 1-liter concept car. Similar to the Very Light Car in some ways—lightweight, low drag—it gets a purported 261 MPG and is headed into small-scale production.

Oliver doesn't think the industry's new fascination with lightness goes far enough. To him, they're still fiddling around the edges. "Look, General Motors today has a really good fleet of cars," he says. "But they're still the same fleet of cars." He pauses. "They have figured out how to lease the space without cutting the window."

As we cross into the hilly region outside Lynchburg, I ask him what he thinks of the Tesla Model S. He says it's a beautiful car. Tesla may end up owning the electric-car market, Oliver says, and "I want them to win." But he's trying to do something different. The moderate sales of mass-market electric cars like the Volt and the Leaf have only strengthened his long-held belief that drivers are waiting for a "leap better car," one both superefficient and inexpensive. And while the automakers may not *want* a car like this, they *need* a car like this, he insists, in order to hit the 2025 regulatory targets.

Later, when I talk to industry expert Daniel Sperling, author of *Two Billion Cars*, he says the automakers are certain they can hit the targets with conventional technology: moderate lightweighting, direct-injection engines, more efficient transmissions, more hybrids. They don't even think they need electric cars to make the numbers. Given that belief, Sperling tells me, there's little chance the big companies would take a risk on something as alien as the Very Light Car. "They just don't need those radical changes." But Oliver thinks the automakers have miscalculated. The way he sees it, they're now engaged in their own X Prize race, six years after a bunch of nobodies started thinking about how to win. He has the jump.

"I don't know if Volkswagen will buy the Very Light Car, but I hope so," Oliver tells me, entering salesman mode: slowing his voice, hypnotically repeating key words, each sentence a brick in the wall of some inevitable future. "And if Volkswagen doesn't buy it, Jaguar buys it. If Jaguar doesn't buy it, Chrysler buys it. If Chrysler doesn't buy it, Mercedes buys it. I am confident somebody will buy it, because I am confident somebody must have it."

He smiles. "I'm actually really glad you're coming right about now. You're gonna like what you see."

IN LYNCHBURG, we pull up to the old jeans factory and walk through the green steel door into the shop. From the second I enter, the space feels more cavernous than I remember. Richard Melos, the machinist, is over near the lathe and the milling machine, and in the mechanics' nook, P.K. and Ron Cerven, formerly of Li-Ion Motors, are immersed in work on a Very Light Car. Brad Jaeger, the young engineer, will arrive later. But certain spaces in the shop are noticeably empty. There's no one at Ron Mathis's desk, near the old-fashioned placard that reads QUALITY ENGINEERING, and no one back in the body shop, where the mechanic Bobby Mouzayck used to work.

I already know the story with Bobby, because I got a call from him a few months ago. Apparently, there was a dispute over a patent. Last year, when Edison2 submitted its patent application for the design of the Very Light Car, the company listed six inventors: Oliver, Barnaby, Ron, Brad, and the two California consultants. Bobby thought his name should also have been on the list. He wanted it for career purposes, to help him get a job in the future—"This *is* my college education," he told me—but mostly he wanted it because he thought he deserved it. After all those hours in the body shop, sweating in the dust, fixing what the engineers had broken, he felt he'd done just as much to determine the shape and style of the car as the listed inventors, if not more. So Bobby talked to a patent attorney about his options. When he discussed it with Oliver, and started quoting patent law, "They're like, 'Oh, you're gonna sue us? Get the fuck out.'" Oliver fired him and

changed the locks on the shop. "Bunch of fuckin' cunts, really,"
Bobby told me, disgusted. (Later, when I ask Oliver what happened
with Bobby, his voice drops to almost a whisper. "Bobby got into
an argument where I couldn't help him," he says somberly. "Bobby
is too volatile. He is not coming back. I feel bad about it.")

Bobby, then, has been gone for months. But Ron's exit is news
to me. In the shop, P.K. wanders over to say hello, and I ask him
about it. "Mathis?" he says, tilting his head. "He's quit twice." He
grins. "Those *sensitive engineers.*"

What's happening here is pretty easy to see. When money
runs low at a start-up, other kinds of currencies become more
important, like attribution, the issue that drove out Bobby. And
if employees can find other work that pays, they take it. Ron has
moved on to new projects, working with multiple race teams,
building a supercar for a guy in the United Arab Emirates. "Money
has been spotty," Oliver says, "and Ron got pissed." Barnaby isn't
working with Edison2 at the moment, either, for the same reason.

I know Oliver too well by now to expect him to interpret
these events as setbacks. Ron will be back, he insists. His plane's
broken, so he can't fly to Lynchburg right now. Some people have
car trouble; Ron has plane trouble. (Ron later rejoins Edison2 in a
part-time capacity.) "Look around," Oliver says. "Most of my guys
are still here."

He's right. The core team remains intact. Even skeptical,
grumbly P.K. has stuck around; despite his complaints about life
at Edison2, P.K. comes into the shop every day and tries to make
the car better. He believes in it, or maybe it's just that he believes
in his boss. The engineers and mechanics still see something in
Oliver. Yes, there are times when he loses focus, or comes on too
strong and scares away potential allies. But there are also times
when he lashes the car to a larger vision, when he makes the work
in the shop feel like a wondrous adventure that will rewire history
because it must, and then the men remember why they've stayed.

The 4.0.

He tells me he has a plan—a new plan to raise funds, a bold plan to turn Edison2 into a real company.

The plan centers on Edison2's in-wheel suspension system, the one that makes the unusual architecture of the car possible. Edison2 will license the suspension to a new company Oliver will create called Edison4. Edison4 will develop and market the suspension to major automakers, to replace existing suspensions in their coupes, sedans, minivans, and SUVs. With an Edison4 suspension, "Any car will weigh 200 to 400 pounds less, cost less, and perform the same," Oliver says. He'll try to grow Edison4 by accepting investments from the public. Meanwhile, Edison2 will continue to optimize the entire car.

Now Oliver directs my attention to the car in progress near P.K. The last time I was in the shop, a year ago, the Very Light Car 4.0 was just a nonfunctioning model made out of foam, but now it's been realized in metal. Oliver says in two weeks it will be a fully functional "mule" car—a prototype that will be driven hard and tested hard and refined along the way. He says it will probably take another three years of development before the completed car is ready, in 2016.

The goal is to make it a car that people will want to drive every day, and also to make it easier to manufacture. The main difference from the X Prize prototype is that the frame has been redesigned to use skins of sheet aluminum instead of a carbon-fiber body wrapped around a steel-tube frame; the skins will be simpler to stamp out of metal. The car is also wider and taller, for better passenger comfort, and the suspension has been overhauled to allow for more wheel travel and to cut down on road noise and vibrations. The version I'm looking at is electric, with a modest-sized battery pack and an electric motor. Because of the battery pack and the sheets of metal, the car has gotten heavier. "Fourteen hundred pounds," Oliver says, not batting an eye. The initial impression of the car is less alienating; the aluminum skins give it a new sense of solidity. The California consultants have produced a set of 3-D renderings of the eventual completed car, color printouts of which are tacked to the shop's wall. In the renderings, the car looks sharky, the hammerhead bumper painted with a futuristic leopard-skin pattern.

Oliver says he's also made progress on safety. After the X Prize, he and his staff crashed a car at a private testing facility in Indiana. He tried a test devised by the Insurance Institute for Highway Safety, a 40 percent "frontal offset" crash, in which only part of the car's front slams a barrier head-on. The crash car's frame is in the other room of the shop. The left front wheel pod is completely shorn off, but otherwise the car appears to have suffered little damage; the steering column is intact. When the car hit the barrier, instead of bouncing back like a normal car, it glanced off to the side. According to a graph of data from the crash, the Very Light Car suffered half the deceleration of a normal vehicle, lending credence to

Edison2's theory that in some kinds of crashes the car will deflect damage like a judo fighter. "It didn't just sort of pass," Brad tells me. "It blew the other cars away." But it's still hard to say how the Very Light Car will fare in a full suite of crash tests.

Before I leave the shop and drive back to Philly, Oliver tells me a couple of things I haven't heard before. One is that he thinks he can make an even simpler version of the Very Light Car for the developing world. The developing-world car "has more noise, more vibration, but it still handles just fine," he says. "Everything that holds a bracket is molded out of plastic, like a McDonald's crate. It's a steel frame with plastic molded around it. The cost of this is nothing. Nobody's ever done it. A piece of plastic, a piece of steel. This is a car that doesn't exist. This is a car that costs as much as a nice motor scooter. This is what is possible if you follow this philosophy."

Oliver's other new obsession is Google's self-driving-car technology. He recently gave an hour-long talk at Google headquarters in California. "Yes, I drove the Google car," he says. "Oh fuck yeah. They like this, they get it. Here: You build the car out of plastic, the way I'm talking. You have a tiny electric battery. It seats two people. You get into the *individual transportation pod*. You get into the Google commuter lane. The bus lane is now the Google lane. And Google drives me to work in my little pod." Oliver tells me that with a Very Light Car piloted by a Google brain, a car becomes even more efficient than a bus, per person. "I don't wait for the bus, I don't have to stand outside, and I do it with less energy? Which one's gonna win?

"Look. This isn't gonna happen tomorrow. But this *could* happen. Now the pieces exist."

He's rubbing his fingers together.

"I can see it. I can feel it. I can smell it."

———

FOUR MONTHS pass before I hear from Edison2 again, in an e-mail that arrives in April: *Edison2 to Unveil New Very Light Car Architecture at the Henry Ford.*

The car must be done. They're ready to show it to the public.

A few days later, I'm in Dearborn, Michigan, inside a hall of freshly waxed old cars. It's part of The Henry Ford museum, founded in 1929 to document what Ford called "the history of our people as written into things their hands made and used." (He originally named it the Edison Institute, in honor of his friend and idol Thomas Edison.) As you'd expect, the museum is full of lavishly restored Fords from the turn of the century—Model A's and Model Ts, the machines that made history. Yet some of the cars are surprisingly mundane: a mauve 1955 Chevy hardtop, a teal 1960 Chevy Corvair, a gray 1949 Volkswagen sedan, a tan 1978 Dodge Omni "econobox." A placard in front of each car gives its price in the currency of its day, as well as its weight. You can see the whole sweep of it here: how cars started out tall like carriages, then became low and streamlined, then boxy, then small, then bigger again, all in response to broader shifts in taste, need, and corporate will. The Henry Ford isn't trying to awe its visitors with gearhead stats. It's trying to tell a story about cars as social objects.

Near a deep-red 1909 Model T is one of the Very Light Cars from the X Prize; Oliver donated it to the museum after the contest. Next to the car, the museum has set up a space for him to give a speech and a press conference. Twenty rows of chairs face a podium and a flat-screen TV. To the right of the podium, there's a car-shaped object obscured beneath white fabric and, behind that, an enormous screen, also covered with a white sheet. Fifty or sixty people are milling around in the area, waiting for the moment. Some are journalists. Some are Oliver's people from Virginia; a few of his personal friends have made the trip, as well as Ron Mathis, Brad, Richard, and P.K. of Edison2. The rest, as far as I can tell, are just museum-goers, curious about what's under the sheets.

Oliver isn't hard to spot. He's the tall guy wearing a jacket with patches on the elbows and a significant hole in one shoulder. I wave, and he walks over and greets me warmly. He says the museum's interest in the practical dimension of cars makes it the perfect place for Edison2's big unveil: "We are about green, reality, price." He leans in conspiratorially. "And we are *killing* price. We are killing it."

I gather that the main objective of today is to find an investor, a partner. Oliver is trying to signal that he has reached an important milestone in the development of the car, three years after the conclusion of the X Prize. He wants to go even further, but he needs to find someone to work with him.

The president of the museum, Patricia Mooradian, takes to the podium. She gives an overview of the Very Light Car's accomplishments in the X Prize. "I can see my future," she says. "One of those is in my garage. How about it, Oliver?"

Mooradian yields to Oliver, and he launches into a pitch that's part paean to his own success and part ode to the potential of the technology itself. "We are not foolish people. We've been at this for five years now. I suspect we'll be at it the rest of our lives. The reason being very simple. This *is* disruptive technology. This can change the entire industry. This can change the economies of nations."

From here, Oliver loses steam, because he has to go back and explain the auto prize and why it mattered. In figuring out how to win the money, he says, Edison2 solved a number of problems. How to make an efficient car that can burn gasoline. How to make a light car that's safe. He starts to ramble. Now he's talking about the weekend he spent with Ron *before* the X Prize, ticking through various alternative fuels and settling on gasoline instead. Now he's getting into the weeds of carbon dioxide emissions. The speech starts to feel like a lecture. Oliver cues a graph on the TV that shows the current emissions of the major automakers' fleets compared with a curve representing future emissions targets; the automakers, right now, are all way above the curve, not below it as they need to be. "As you can see, the industry is a long way away."

For the last three years, he says, Edison2 has been trying to turn its X Prize prototype into something more than a science project, "something that meets people's expectations." He walks over to the screen covered with the white sheet, and with the help of the museum president, pulls the sheet away, revealing a full-color rendering of the finished Very Light Car 4.0. Then Oliver removes the sheet from the car itself. It's the 4.0 in skeleton form: metal frame, wheels, tires, seats, but no body. The moment is nothing like a big unveil at a car show: no camera flashes, no pumping music, no oohs and ahhs from the crowd, just Oliver in his holey suit.

Now Oliver begins to discuss the 4.0's new features: the added girth, the smoother ride, the trunk space. He goes on for fifteen minutes. Then he opens the floor to questions.

A man asks when the public will be able to buy a Very Light Car. Oliver replies, "When it's ready and it's really *good*."

There aren't many other questions. Oliver invites the audience to come have a closer look. He smiles and backs away from the podium. The audience applauds. A dozen people walk up and cluster around the car, taking turns sitting in the driver's seat.

I don't know if Oliver will get what he needs from this event, or if Edison2 will survive. I only know that he can't be killed. Bomb everything to rubble and watch him gather scrap. The Internet goes dark and he lights a match. And not him alone, but all his ingenious kin: every kid in a shop program, every girl and guy in a garage, every hacker and maker with no hope of bailout by bank or by nation.

He's standing next to the car now. He's watching the journalists circle it and circle it and write in their notepads. He reaches down with both hands. He grabs a metal bar. He pulls up, gently, eyes lit with joy, and the car rises into the air.

acknowledgments

THIS BOOK wouldn't have been possible without the efforts of many, starting with my editor, Rachel Klayman. This project was supposed to take a year and a half. It ended up taking three and a half. But Rachel stuck with me. She encouraged my better tendencies and saved me from my worst, and the book is immeasurably sharper for it. At Crown, Stephanie Chan and Emma Berry made helpful suggestions, and Molly Stern was a staunch ally. I also appreciate the beautiful work of designer Barbara Sturman and cover designer Jamie Keenan, and I'm lucky that Rachel Berkowitz, Cindy Berman, Catherine Cullen, Jacob Lewis, Matthew Martin, Jessica Prudhomme, Annsley Rosner, Courtney Snyder, and Jay Sones were there to shepherd the book out into the world.

My agent, Larry Weissman, and his partner, Sascha Alper, have a fingertip feel for what makes a good story and how to shape an idea into a book. I've relied on their judgment more times than I can count. They are brave, funny, passionate, and tireless, and I feel humbled by their support, which lets me make a living doing what I love.

There was a point early on when I was having trouble finding the story, and I reached out to a writer friend for help. Kari Walgran filled the margins of my initial drafts with sharp, witty observations, as well as jokes that made me laugh out loud. When I was lost, Kari showed me the way into the tale. I could not have done it without her. The same goes for Alex and Susan Heard, whose careful scrutiny of a later draft helped me make numerous improvements, and for Jay Sullivan and Marilyn Moyer, who were constant friends and sounding boards. Several talented colleagues either read drafts or talked out problems with me, including Dana Bauer, Steve Volk, Matthew Teague, Brendan I. Koerner, Justin Heckert, Matthew Shaer, Andrew Pantazi, Caleb Hannan, Michael Mechanic, John Voelcker, Seth Fletcher, Christopher McDougall, Stephen Rodrick, and Sasha Issenberg.

I also appreciate the support of the magazines that let me write for them. I'm lucky to have editors like Peter Rubin, Bill Wasik, and Chris Baker at *Wired*; Tom McGrath and Michael Callahan at *Philadelphia* magazine; Mark Healy at *Men's Journal*; and Brendan Vaughan at *GQ*.

I sent portions of the manuscript to several sources to review for factual accuracy, on the condition that I would make the final decision about every line of text. Oliver Kuttner, Brad Jaeger, Ron Mathis, Barnaby Wainfan, Bobby Mouzayck, David Brown, Kevin Smith, Jen Danzinger, Ann Cohen, and Simon Hauger read the material carefully and provided important clarifications. Any errors are mine alone.

I want to thank the people who made my reporting possible. Peter Diamandis, Cristin Dorgelo, Eric Cahill, Eileen Bartholomew, John Shore, Will Pomerantz, Vijay Goel, and Chelsea Sexton contributed key insights into the workings of the X Prize Foundation. At the track in Michigan, Julie Zona, Steve Wesoloski, Jody Nelson, Bob Larsen, and Valerie Arias were kind and knowledgeable hosts, and the excellent automotive journalist Ronald Ahrens let me pick his brain. Throughout the process, the capable team from C. Fox Communications—Carrie Fox, Arron Neal, Lauren Reese, and Melissa DeLaney—made sure that I had the access I needed. I'm also grateful to Greg Fadler and Suzy Cody of General Motors, who couldn't have been nicer, more forthcoming, or more obviously devoted to their craft.

For their love, support, babysitting shifts, and emergency bridge loans, thank you to my parents, Frank and Sharyn Fagone; my sister, Lauren Fagone; and my in-laws, Lynn and Rich Bauer.

Thank you to Dana, my wife, and Mia, my daughter, for being wonderful.

Finally, the competitors. At first, it must have been obvious to them that I had more enthusiasm than knowledge when it came to efficient cars. Still, Oliver of Edison2, Simon of West Philly, Kevin and Jen of Illuminati, and Marques McCammon and Tom Reichenbach of Aptera opened their shops to me, answering endless questions with patience and charm. I'm grateful for their trust. I also owe much to their remarkable collaborators: at Edison2, Ron, Brad, Barnaby, Bobby, Peter "P.K." Kaczmar, Reg Schmeiss, Julian Calvet, and David Brown; at West Philly, Ann, Kathleen Radebaugh, Jerry DiLossi, Ron Preiss, Jacques Wells, Sowande Gay, and Azeem Hill; and at Illuminati, Nate Knappenburger, Josh Spradlin, Thomas Pasko, George Kennedy, and Nick Smith. To all of the inventors, hackers, teachers, students, engineers, mechanics, and entrepreneurs who shared their stories with me, thank you.

About the Author

JASON FAGONE has written about science, sports, and culture for *GQ, Wired, The Atlantic, New York* magazine, *Grantland, Mother Jones, The New York Times Magazine,* and *The Best American Sports Writing.* He is the author of one previous nonfiction book, *Horsemen of the Esophagus,* about competitive eaters, and is the recipient of a Knight-Wallace Fellowship in journalism. He lives in Texas with his wife and daughter. Visit @jfagone on Twitter